# Alternative Energy Projects for the 1990s

## John A. Kuecken

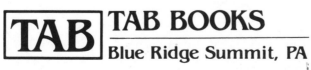

### TAB BOOKS
Blue Ridge Summit, PA

FIRST EDITION
FIRST PRINTING

**Library of Congress Cataloging-in-Publication Data**

Kuecken, John A.
   Alternative energy projects for the 1990s / by John A. Kuecken.
     p.    cm.
   Includes index.
   ISBN 0-8306-8835-8 (h)    ISBN 0-8306-2835-5 (p)
   1. Energy development.  2. Renewable energy sources.  I. Title.
TJ163.2.K84  1991
621.042—dc20                        91-20557
                                                CIP

TAB Books offers software for sale. For information and a catalog, please
contact TAB Software Department, Blue Ridge Summit, PA 17294-0850.

Acquisitions Editor: Kimberly Tabor
Book Editor: Peter Sandler
Production: Katherine G. Brown
Book Design: Jaclyn J. Boone

# Contents

# Introduction

THE INVENTION OF THE WATERWHEEL AND THE WINDMILL ARE LOST IN antiquity. We know these devices were widely used in the middle ages; however, the wide use of the energy stored in combustible matter did not really begin until the development of the atmospheric steam engine by Newcomen in 1712 and the improvement of the steam engine by Watt in 1769. By 1815 Oliver Evans of Philadelphia had developed a high-pressure steam engine operating at 200 pounds per square inch (psi). By the 1820s most major laboratories and some advanced manufacturing facilities were equipped with steam engines and line-shafts and belt drives. This was the beginning of the inexorable drive to replace the muscle power of men and animals with the tireless efforts of the machine.

The steam engine, the waterwheel, and the windmill were not to be the entire answer. Each of these machines was large, cumbersome, and expensive, and the power that it developed could not be transported over any significant distance without great expense and losses. Yet another development was required. This came at first in the form of the demonstration of the dynamo-electric machine by Michael Faraday in 1831. Faraday demonstrated that electrical energy could be turned into mechanical energy and that an invisible electric current could also be made to rotate a shaft at high speed like a steam engine or a waterwheel. A Frenchman by the name of Hypolite Pixii invented the machine in 1832 that was eventually to turn the course of history. The Pixii machine was a heteropolar permanent magnet commutated machine with an iron armature and a wire-wound core. It provided the first means to generate a continuous current from mechanical energy and freed the electrical experimenter from the requirement to generate electricity by burning expensive fuels like zinc, copper, iron, and sulphuric acid in a chemical reaction.

The major significance of this was pointed out by Professor Siemens in 1851. At the speeds possible in those days, a one-inch diameter shaft could transmit perhaps one horsepower with a bearing every three feet. In less than a mile, the entire power would have been consumed in bearing friction, even with the finest bearings available. By comparison, the same one-inch diameter shaft made from copper could conduct over 2000 amperes of electricity. At 115 volts, this conductor could transmit 363 horsepower for a mile with very modest losses. In modern transmission line service at 220 kv, this same conductor would handle 695,000 horsepower. There is no torque; therefore, no bearings are required, and the conductor can be suspended from widespread poles or towers with only insulators to suspend it.

It was this ability to transmit energy easily and economically over great distances that led eventually to the near elimination of physical drudgery by man and beast in most of the advanced nations of the world. In 1884 Edison opened the Pearl Street station in New York to sell electrical energy to operate the incandescent electric lamps he had invented and was then manufacturing. And the first central electrical power generating facility went "on line."

# The diffuse energy system

For a period of nearly 85 years following the opening of the Pearl Street station, the consumption of energy and the size and power of the energy companies increased at a rate of nearly 8 percent per year, doubling about every 9 years with little real concern expressed by the public. Assuredly, there were periodic voices raised to note that the supply of fossil fuel was limited. As a matter of fact, a review of back issues of *Scientific American* reveals that about every other year some professor would calculate the year in which we would run out of coal or petroleum based on current rates of consumption, growth, and known reserves. However, the public is not prone to worry about such matters and the expansion continued unabated.

The energy companies were less sanguine about the subject and began to push into nuclear energy as an answer to long-term requirements for energy growth. This move has at times met with vigorous opposition from groups of people, some of whom oppose any nuclear installation with a ferocity that approaches religious zeal.

At the same time, the depredations wrought upon the state of our planet by the continuing consumption of energy in the form of fossil fuels or nuclear fuels has brought about a substantial public sentiment for the utilization of renewable resources for the satisfaction of our continually growing thirst for energy.

There are only two fundamental sources of energy available to mankind: the nuclear energy locked into certain elements in the death throes of long-vanished stars and the radiant energy from the nuclear fires of our own sun. The latter is bequeathed to us in the form of the fossil remains of long-dead forests and in today's wind, rain, and sunshine. Arab oil embargoes and coal strikes have taught us that the fossil fuels are a fallible source of energy. Indeed, the blackouts have taught us that the energy giants are also fallible.

At the time of the writing of the first edition of this text in 1978 there was a widespread feeling in some circles that the future lay with a diffuse energy system in which every home would have some mechanism for generating its own energy.

The homes would be warmed by sunshine and electrical requirements would be satisfied with some form of self-renewing or renewable resource. There were a host of start-up firms that intended to ride the wave of energy self-sufficiency to success and happiness.

Thirteen years later, it is more or less obvious that the tremendous wave of energy self-sufficiency did not materialize; at least not in a vast wave of commercial success. Among those who have commercial power conveniently at hand, there are precious few who are not connected and who are running entirely on energy they generate or collect for themselves. One scarcely hears of "The Diffuse Energy System" anymore.

On the other hand, some relatively subtle changes have taken place. A significant fraction of the home owners in the colder climates have added insulation to their homes. Residents in the warmer climates have also to reduce the air-conditioning load. More energy-efficient furnaces, hot water heaters, air conditioners, and refrigerators and freezers have made their appearance. Furthermore, people are willing to pay a premium for items with efficiency advantages. Federal law requires statements of estimated energy consumption on such appliances and salesmen will sell by them. In short, the public has become conscious at least of the cost of energy.

The results of these changes have been felt by the energy companies. The growth in usage of electric power is down from 8% per year to 2% in the Northeast and East. There are some factors that make the interpretation of such figures a bit uncertain; one of these is the decline in the birth rate. The baby boomers are marrying later and having fewer children. Another factor is the exodus to the sunbelt. The East, Northeast, and Midwest have tended to slow down to very small growth figures and in some areas population is on the decline. However, the slowing trend is clearly distinguishable in all energy areas including fuel oil, natural gas and gasoline.

On the commercial side a little progress has been made with renewable energy. In the January 1986 IEEE Spectrum (p. 68) it is reported that 500 million kilowatt hours of electricity was commercially generated in 1985. Of that 99% was generated in California. This would have consumed 80,000 barrels of oil and would have added some quantity of pollutants to the atmosphere. However, to place this in perspective, the entire 1985 commercial wind output is the equivalent of 40 minutes of full output from Three Mile Island 1, which went back into operation Oct. 9, 1985.

During 1984 the installed wattage for photovoltaic plants rose to 18 megawatts and Solar Thermal accounted for 28 megawatts. By comparison, the delivered energy for more conventional power sources in the US was:

| | |
|---|---|
| Coal | $1,350 \times 10^9$ 9 wh |
| Hydro | $300 \times 10^9$ 9 |
| Natural gas | $280 \times 10^9$ 9 |
| Nuclear | $305 \times 10^9$ 9 |
| Petroleum | $80 \times 10^9$ 9 |
| Total: | $2,315 \times 10^9$ watt-hours. |

# The purpose of this book

The subject of the energy industry in the US is fraught with political opinions, and this is not intended to be a political book. Instead, it is intended as a practical book to present the pros and cons of various home energy producing schemes. As a bow to practicality, the book deals with the economic aspects of each approach to the extent possible. Computer programs capable of running on either Apple II+ (c or e) or IBM and other relevant computers are presented to compute the real cost of financing an installation. Programs are also presented to permit you to calculate how much investment is warranted for a known savings per month and a Return-on-Investment program is presented to calculate how long it will take you to get your money back. The programs include income tax considerations.

Another bow to practicality is the recommendation of junkyard parts when their use is practical. A number of places where the use of used or scrap items will save money are noted.

With a final note on the subject of practicality, the main emphasis in the book rests on generating 115-volt 60-Hz alternating current. Low voltage dc is easier to make and easier to store; however, very few of us have any significant appliances that will run on anything but 60-Hz electricity. There are refrigerators and other items made to run on low voltage dc for use on yachts or perhaps recreational vehicles; however, these items are relatively expensive and difficult to service. You will not find them on sale at your friendly appliance store.

One final note is in order. Some of the concepts involved in establishing a home electrical system require some knowledge of electricity. If you do not have some grasp of elementary electricity, a little study in this area is prerequisite. Also, it will usually not be possible simply to buy a turnkey installation. In general you will have to do at least some designing on your system. It is probably not possible to "cookbook" a system. This means that some algebra is involved in the system design. Wherever possible, I have tried to give several clear, simple examples and perhaps a computer program to make the math as painless as possible. For those not disposed to work on the system design, I strongly recommend that you contract with a professional to design the system.

A private electric system is too small to get by without some careful planning on load control and distribution.

Let me add that, I hope you will have as much fun reading this text as I did researching and writing it.

# 1

# Home electrical requirements

A LOGICAL PLACE TO START THIS BOOK IS WITH A CONSIDERATION OF JUST how much electricity one is likely to need in a home. If one is seriously considering the installation of an independent electrical plant, the proper sizing of the installation is of considerable importance. If a person wants only enough electricity to operate a transistor radio or perhaps an emergency lamp, the solution can be achieved very simply with a few solar cells and a storage battery. However, if you intend to operate an electric automatic drier and an air conditioner along with the usual household appliances, a considerably more complex and expensive installation is required.

A fair place to begin, assuming that your home is now electrified, is to survey your electric bills. The power company prints the number of kilowatt hours on the bill. You will find that the examination of the year's bills will show a certain amount of seasonal fluctuation so it is best to consider the billing for the entire year. I live in a fairly typical suburban home in a cold northern climate and at 47° North latitude. The days get relatively short in December, so the lights are burned more often and for longer periods. In addition, the furnace blower runs more frequently. Added to the usual lights we have the following appliances:

- Small refrigerator
- Large self-defrosting refrigerator/freezer
- 22-cu. ft. freezer
- Steam iron
- Automatic dishwasher
- Automatic clothes washer
- Automatic electric dryer
- Gas furnace with electric blower

*continued.*

- Electric stove
- Small electric oven
- Disposal
- Hair dryer
- Large color TV
- Small color TV
- Stereo (solid-state)
- The usual assortment of radios, mixers, etc.

We do not have an air conditioner and do not heat water electrically. However, this is somewhat offset by the presence of a fairly extensive electronics lab and machine shop. The home is generally occupied by two people during the college year, and the population can rise to six or so during holidays.

In a typical year I find that we consume about 10,200 kilowatt-hours (thousands of watt-hours). Because there are 8766 hours in the average year, this works out to an average electric load of 1.164 kilowatts. The peak month is February, the lowest is June. At a cost of 8.5¢/kWh this works out to $867/year. In the next chapter we will deal in greater depth with the cost.

# Average power loads

The yearly average power found on the power company bills tells only a part of the story, and that part is principally economic. The monthly peak power is generally reached in February, which is normally the coldest month in our northern climate. During this period, power consumption rises to about 960 kWh/month or an average load of 1.314 kW. During February, the solar input is minimal, and it might be difficult to keep a small hydro plant operating since small bodies of water will freeze to a depth of a meter or more. A wind-operated plant would have a better chance for success provided it could be kept from icing up.

Naturally, these comments do not apply to locations like Tucson, AZ, or the Florida Keys. However, the point remains the same: The wind does not always blow, the water might not always flow, and the sun certainly does not always shine, at least not at night. In order for a home electric system to be completely self-sufficient, it must be provided with enough capacity to store energy to carry at least a reduced load during periods when the "free" prime energy resource temporarily disappears. In certain climes it might be advisable to provide alternate access to prime energy with both a wind-and-water or a wind-and-solar installation or some other combination. For example, a solar steam installation could be built to include a conventional boiler that would burn some renewable-resource fuel such as wood or cornstalks. This action does carry the burden of introducing combustion products into the atmosphere.

To put the matter into perspective, a very efficient gasoline engine consumes 0.6 pounds of fuel per horsepower-hour. There are 745.7 watts in one horsepower, so our 1.3 kW represents 1.76 horsepower. A small wood-burning steam plant would be no more than one-fifth as efficient and could easily consume 5 to 10 times this weight in wood or cornstalks or other low-grade fuel per horsepower-hour. This translates into a fuel consumption of between 5¼ and 10½ pounds of fuel per hour to maintain our 1.3 kW electrical load. One month without any significant sunshine would require the consumption of between 3800 and 7700 pounds of low-grade fuel. With typical hardwoods, this would represent a woodpile in a cube five feet to seven feet on a side.

A side benefit of a solar steam system would be the fact that the "waste" heat from the condenser would be capable of providing a good share of the heating requirements of a rather large house. This would be true regardless of whether the system was operating on sunshine or the auxiliary fuel. The "waste" heat capability is an advantage not accruing to either a falling-water or a wind-driven plant and would tend to offset the naturally higher cost of a solar steam installation. In our discussion of solar steam installations, we will examine this facet in more detail.

In a hot, sunny clime such as the American Southwest, the outlook for a solar steam installation for electrical energy is much more attractive. First of all, the sun shines more often. Secondly, the peak system load will generally be due to air-conditioning, which is needed primarily when the sun is shining. In many of the desert locations the night air is rather cool, and little if any air-conditioning is required at night. In such locations, the principal nighttime loads can probably be confined to food refrigeration and lighting.

In any event the maximum average load, coupled with an estimate of the duration for which this load must be maintained, provides a measure of the energy storage capability that must be incorporated into the system. Probably the only system that could reliably operate without some energy storage is a hydro installation in an area where the water supply rarely dries up or freezes.

# System sizing

The average power draw actually is only one of the factors involved in a small electrical plant. The small plant size is more nearly related to the maximum load that is to be handled. A reasonable explanation of this is involved with some of the properties of alternating current. There is no way to describe these properties without getting a little mathematical and one is not going to be able to understand what is happening with the system unless he has some grasp of these properties.

For example, in a dc circuit, the power is simply the product of volts times amperes; in an ac circuit this is not true! In a dc circuit where a group of wires form a junction, the sum of the currents in the junction is zero; that is, the currents leaving exactly cancel the currents entering. This is also not true in an ac circuit. In a series dc circuit, all voltages within the circuit are less than the voltage across the input terminals; not so for ac. The voltages and currents within an ac circuit can be many times larger than those at the entrance terminals; lethally so in some cases!

Unless you intend to have someone design and install a "turnkey" system for you and unless you intend to pay for a system that is much larger than absolutely necessary, it behooves you to have some grasp of a few of the fundamentals.

# Phase angle

In order to take some of the curse off of the algebra, I have written some computer programs to illustrate and in some cases to solve the practical problems. The illustrative programs are intended to help the reader visualize some of the concepts that seem to give people trouble.

The first concept we deal with is the concept of phase angle. Suppose that I took a yardstick and nailed one end to the wall. I could swing the free end in a circle. The height of the free end above (and below) the nail would be 36·sin(th) inches; where (th) is the angle between the horizontal and the stick itself. It is very easy to see the fact that the free end goes above and below but the shape of the height/time curve is less easy for some folks to visualize. Program 1-1 will run on an Apple II+, e or c, and illustrate both the rotating yardstick and the plot of height versus time. Figure 1-1 shows the screen after slightly more than one revolution. The oldest part of the sine wave is to the right.

**Program 1-1. Sine wave demo routine, by Jack Kuecken**

```
100    REM ::::::::::::::::::::::::::::::::::::::::::::::::::
105    REM        SINEWAVE DEMO ROUTINE
115    REM         BY JACK KUECKEN
120    REM            01/16/86
125    REM ::::::::::::::::::::::::::::::::::::::::::::::::::
130    HGR2
150    POKE  - 12524,0
160    POKE  - 12525,64
170    POKE  - 12529,255
180 X0 = 45:Y0 = 95
190 R = 44
200    HOME
210    DIM Y(280)
212    FOR I = 280 TO 90 STEP  - 2
214 Y(I) = 95
216    NEXT
225 TH = 3.14159265 / 2
240    REM *******************************************
242    REM        THE RADIUS ROUTINE
250 Y = Y0 -  (R *  COS (TH))
260 X = X0 +  (R *  SIN (TH))
270    HCOLOR= 0
272    REM  PLOT BLACK
280    HPLOT X0,Y0 TO X1,Y1
290    HCOLOR= 3
292    REM  PLOT WHITE
300    HPLOT X0,Y0 TO X,Y
310 TH = TH -  .1
320 X1 = X:Y1 = Y
400    REM *******************************************
410    REM        THE SINEWAVE ROUTINE
420    HCOLOR= 0
422    REM  PLOT BLACK TO ERASE THE SINEWAVE
```

```
430   FOR I = 278 TO 90 STEP  - 2
440 J = I - 2
450   HPLOT I,Y(I) TO J,Y(J)
470 Y(I) = Y(J)
472   REM   MOVE THE POINTS OVER
480   NEXT
490 Y(90) = Y
492   REM  GET NEXT PROJECTION
540   HCOLOR= 3
550   FOR I = 268 TO 92 STEP  - 2
560 J = I - 2
570   HPLOT I,Y(I) TO J,Y(J)
580   NEXT I
590   GOTO 250
```

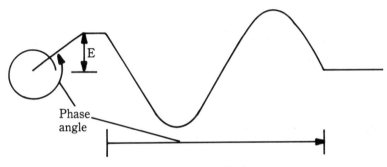

**1-1**  Rotating vector display.

Lines 130 thru 170 simply tell the computer to plot on a high resolution page, print black on white, print out high res page 2, and operate the printer unidirectionally. These instructions work for the Silentype. Other printers might ignore them or not require them.

Line 180 establishes the location of the nail and line 190 establishes the scaled length of the yardstick. The rest of the program has to do with plotting the yardstick and the locus of the height of the tip. Replotting with HCOLOR=0 erases what was just plotted when HCOLOR=3. Line 470 is the thing that makes the sine wave creep to the right.

The voltage at the wall outlet is 163•sin(377•t) (if you live in the US). T is measured in seconds and the angle is in radians. It is actually helpful to think of the voltage vector spinning merrily away and the phase angle winding out.

In actual electrical work we are seldom interested in the absolute phase angle as given above because this increases with time quite rapidly. In general we are interested only in the difference in phase between voltage and current or between two voltages or two currents. The routine of Program 1-2 plots the locus of voltage and current versus time after having you assign a phase difference in degrees. The Apple does not understand degrees, so line 145 converts to radians. It also does not know the value of PI; therefore line 140.

**Program 1-2. Sine wave power routine, by Jack Kuecken**

```
100   REM ::::::::::::::::::::::::::::::::::::::::::::::::::
105   REM          SINEWAVE POWER ROUTINE
115   REM            BY JACK KUECKEN
120   REM               01/16/86
125   REM ::::::::::::::::::::::::::::::::::::::::::::::::::
126   HOME
127   INPUT "ENTER PHASE ANGLE IN DEGREES";TH
128   IF TH > 90 OR TH <  - 90 THEN  PRINT "IT MUST BE BETWEEN +
      -90": GOTO 127
130   HGR2 : HCOLOR= 3
140 PI = 3.14159265
145 TH = TH * PI / 180
150   POKE  - 12524,0
160   POKE  - 12525,64
170   POKE  - 12529,255
180 YO = 95:R = 90
190   REM ****************************************
195   REM          THE COMPUTATION
200 A2 = A1 + TH
210 E = R *  COS (A1)
220 I = R *  COS (A2)
230 P = E * I
240 PT = PT + P / (R * R)
250   REM  STORE THE ENERGY
290   REM ****************************************
300   REM          THE PLOT ROUTINE
310 Y1 = 95 - E
320 Y2 = 95 - I
330 Y3 = 95 - P / R
340   HPLOT X,Y1
350   HPLOT X,Y2
360   HPLOT X,95 TO X,Y3
370 X = X + 1
380 A1 = A1 + .04488
390   IF X < 280 GOTO 200
420 PT = PT / 280
600   STOP
```

You will remember that I said the power in an ac circuit was not simply V·I. Well, that is true in the long haul but at any instant the power is simply V·I; however, this is true ONLY if one keeps track of the "negative" power generated when V is plus and I is minus, and vice versa. This concept might offend some of my more formal colleagues, but it will suffice for our purposes here.

Figure 1-2 shows the views for TH = 0 and TH = 15 deg. At zero, the current and voltage are simply plotted one atop the other and only a single sine wave shows. In the lower illustration the current does everything 15 degrees sooner than the voltage. The curves are otherwise identical. If we skip ahead to Figs. 1-3 and 1-4 and look at the lower illustrations we see that there is beginning to be a considerable amount of "negative" power. Now going back to the program, we see that

PRINT PT
.500001168

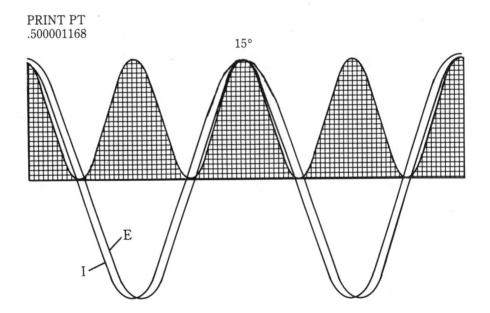

]PRINT PT
.482964055

**1-2**   Power as a function of phase angle.

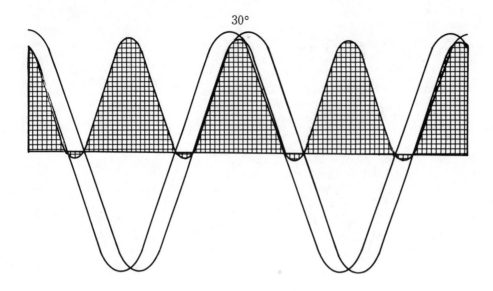

```
]PRINT PT
.43301374

]
```

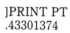

```
]PRINT PT
.353554255
```

**1-3**  Power as a function of phase angle.

line 240 normalizes P to 1 volt and 1 amp and saves the sample in an accumulator PT. Since we know that there are 280 samples plotted (the width of the Apple screen) we can divide by 280 on line 240 to compute the average value of P. At zero phase shift we find that PT = 0.5 and at 90 degrees phase shift we see that PT = 0 (well, close enough for government work). As a matter of fact, you can take your pocket calculator and check to see that PT = 0.5•cos(th). This becomes in the general case:

$$WATTS = VOLTS \cdot AMPERES \cdot \cos(TH)$$

In other words, we have to know the phase angle between the voltage and current in order to calculate the power.

### RMS Values

Before we leave the program entirely, another point can be made. For a dc circuit with 1 volt input and 1 ampere of current the power would be 1 watt. In the ac case with 1-volt peak and 1-ampere peak and cos(0) = 1 we obtain 0.5 watts average or half as much average power. In other words, in terms of heating effect, the effective values of V and I are below the peak values. Since the effective power is down by a factor of two, and the same factor must alter both V and I, each of these is diminished by the reciprocal of the square root of two. You might have wondered when I noted that the line voltage is given by 163•sin(377•t); however, 163•1/(sqr(2)) = 115.25 volts.

In other words, an ac sine wave with a peak amplitude of 163 volts will heat a resistor the same amount as 115.25 volts of dc would. The heating value is called the Root Mean Square value (this is what we did in the program). The line voltage is usually quoted by RMS value rather than peak amplitude. RMS values are usually given on voltmeters and ammeters. On an oscilloscope, on the other hand, it is more convenient to read peak voltages and currents. Values thus read are multiplied by 0.7071 to convert to RMS.

# Real and reactive power

You may well ask where the energy *really* goes when the phase angle is +–90 degrees. In this condition we can have lots of voltage and lots of current but no power. The answer is that the circuit simply stores and gives back the energy that appears to flow. For example, consider a transformer with no load on the output. As the voltage rises it causes the current to increase in the windings. This current magnetizes the core but the rising magnetism induces a voltage that bucks the current so the current lags the voltage. If the transformer is unloaded and lossless, the current lags by 90 degrees. When the voltage begins to fall, the collapse of the magnetic field induces a voltage that attempts to keep the current up and the current lags by 90 degrees. The same scenario takes place on the negative side of the cycle.

The net result is that the (lossless) transformer simply gives back everything it has accepted from the generator a quarter of a cycle later. This property is called inductance and a device that has it is termed an inductor.

]PRINT PT
.250000631

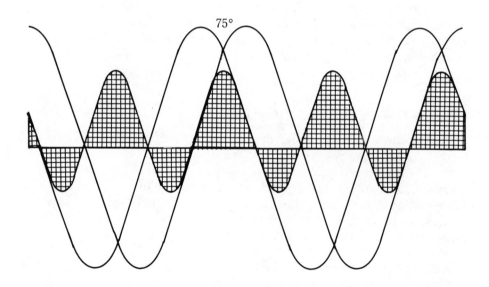

]PRINT
.129409877

**1-4**    Power as a function of phase angle.

The capacitor is more or less the other side of the coin. Whereas an inductor is a kinetic or motion energy storing mechanism, a capacitor is a potential energy storing device. In mechanical terms a moving hammerhead or a moving car has stored kinetic energy whereas a compressed spring or a car parked on a hilltop has stored potential energy.

The voltage on a stored capacitor is a function of the current that has flowed into the capacitor times the length of time the current has flowed. When the voltage crosses the zero line the current flows rapidly into the capacitor since it has no charge to oppose it. By the time the voltage rises to its peak level, the capacitor is fully charged and ceases to draw current. The instant that the voltage turns downward, the capacitor begins to discharge and the current flows in the opposite direction. This is the behavior shown in Fig. 1-5, where the current is leading the voltage by 90 degrees. The capacitor actually gives back all of the energy it stored a quarter cycle ahead of time.

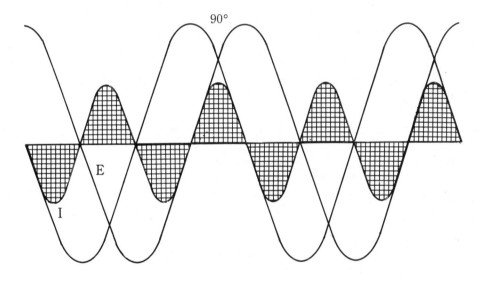

]PRINT PT
5.42202283E-08

**1-5**  Power as a function of phase angle.

Of course, real inductors have some losses, and the current lags by something less than 90 degrees, and real capacitors lead by something less than 90 degrees also. The energy sloshing around in the circuit is usually termed Volt-Amperes-Reactive or VARs to distinguish it from real power.

Now the problem with having a lot of VARs sloshing around in your system is that the current is just as real as the current that provides real power. It will heat your generator and blow fuses just like the current that supplies power to a toaster or incandescent lamp. The numeric value of cos(th) is termed the Power Factor. For a small electric plant the rating is given in thousands of volt-amperes and it makes no difference whether the VA's are VARs or watts. You wind up with your system overloaded even though it is doing little of the work you intended for it.

It is important to note that the watt-hour meter installed in your home is corrected for the cos(th) term and reads actual watts. The power company does not penalize the home owner for a poor power factor, although it does penalize industrial firms. It also does not penalize the home owner for peak power loads, whereas it does penalize an industrial firm.

Since the watt-hour meter is compensated for power factor and the power factor is rarely unity, the current is almost always larger than would be computed from the rate of rotation of the disc.

A lightly loaded induction motor can have a phase angle of up to 75°. Because the cosine of 75° = 0.269, the actual current being drawn from the line is 3.864 times as great as watts divided by voltage. The generator naturally must supply both the voltage and the current even though very little actual power is being consumed. It is this very short-term voltage/current rating that determines the size of any alternating current machine and the size of the prime mover required to drive it. For this reason you will find that a typical alternating current generator will be rated in thousands of volt-amperes (kVA) rather than kilowatts (kW). The plant kVA rating will be a number that you will have to learn in order to construct or specify the right size for your installation.

A rough top limit for your electrical system can be obtained simply by looking at the main fuses in your fuse box. My fuse box contains a pair of 60-ampere fuses through which all the electricity used in the house must pass. The system is a 220-volt split-phase system with 110 volts each side of ground. Figure 1-6 shows this arrangement. Since neither of these fuses has ever blown in a number of years of service, it is safe to assume that the system has never drawn currents in excess of the capabilities of these fuses. The fuses are rated to carry 60 amperes continuously and 120 amperes for one minute. Thus, it is safe to assume that we have never continuously exceeded 220 volts times 60 amperes = 13.2 kVA nor

**1-6**   Electrical service installation.

exceeded 26.4 kVA for a period of one minute. Obviously this is a very large rating compared to the 1.3 kW averaged in the peak month of February.

The reason for this large safety factor lies mainly in the starting characteristics of induction motors. A number of the appliances around the house, such as the furnace blower, the electric clothes dryer, and the automatic clothes washer, will draw currents in excess of 25 amperes when they start. Once up to speed, the current will fall back to something more like five to seven amperes. However, nothing in the existing system prevents five or six such machines from starting simultaneously. In this case the rating of the main fuses is approached or even momentarily exceeded. The power company also has to cope with this problem of transient starting loads. However, they have a very large number of customers, and the probability that all the motors they serve will start at the same instant is negligible except in the fire-up operation after an interruption of service. In order to avoid having something like a 5:1 or 10:1 ratio between installed kVA rating and average power sale, they do two things:

- After any system shutdown, the system is brought up one section at a time and the next section is not switched in until the first has been stabilized.
- The power company protects itself with *power factor correction* which minimizes the value of the phase angle seen at the installation of each section. This power factor correction provides for the minimum number of delivered amperes required to provide the actual load wattage.

In order to provide anything like a reasonable and economic ratio between average electrical usage and installed kVA rating for the plant, a few measures must be taken to ensure some of the same capabilities in a home electrical system. Obviously certain measures could be used to prevent all the motors from starting simultaneously. With available electronics, it is much less expensive to provide a sophisticated control than it is to construct a system with 5 or 10 times the installed kVA ratings. We will also eventually discuss systems that are tolerant to short duration overloads.

The overload problem is especially pernicious in a small system since some systems tend to be severely slow. If the overload occurs because two motors are trying to start simultaneously, for example the refrigerator and the furnace blower, the overload is likely to reduce the system voltage to a level where neither machine can come off of the starting winding and start. In this case the motors will draw a very large current and will usually burn out within a minute or so if not protected by a thermal cutout or fuse. The fuse box fuse will seldom protect a stalled motor from burnout since the starting winding will often burn out if it carries a current not much more than the normal operating current in the run winding. To avoid this type of failure, a great many appliances have a built-in thermal cutout, which opens when the windings start to get hot. This cutout is usually not built to withstand frequent usage since it is only an emergency facility. It is far better to provide your system with some other form of protection against overload and undervoltage.

# Measuring system requirements

If one were starting entirely from scratch to build a home and to buy every one of the appliances new, it might be possible to obtain complete data on the starting and running currents for each of the appliances and do the entire system design on paper. One would have to contact the appliance manufacturer's engineering department and perhaps forgo purchasing appliances for which data were not furnished. However, it is highly unlikely that the salesman in a discount appliance store will be able to furnish you with this information, or for that matter even be able to understand the question. It therefore seems that the home electrifier will have to perform some measurements for himself in most cases.

The illustration in Fig. 1-7 shows three common methods of measuring the ac line current. A standard ac ammeter that connects in series with the lines can be purchased from any number of electrical suppliers. Somewhat easier to use in some cases is the clamp-on ammeter, which is simply clamped over insulated wires. Also shown is the current-transformer/ac voltmeter technique. Note that the measurement is made on the "hot" side of the line.

**1-7**   Three types of ammeters.

**Caution**   Always pull the fuse before connecting an ammeter in series with the line. Never connect an ammeter across the line or between points that show any significant voltage drop.

Failure to observe the first caution will get you shocked, burned, or electrocuted. Failure to observe the second will instantaneously destroy the instrument in a ball of fire.

It should be noted that most modern homes are wired with composite cable carrying the "hot" wire (usually black), a ground return wire (usually white), and a safety ground wire, which can be either green or bare. In order to obtain a reading with the clamp-on ammeter it is necessary to clamp it over *only* the black wire. If all three wires pass through the clamp window, the reading will be zero regardless of the current. The clamp-on ammeter may be safely clamped on an energized wire *only* if the insulation is intact and there are no exposed terminals or bus bars in the vicinity.

Both the standard ammeter and the current transformer require that the circuit be de-energized by removing the main fuses. The "hot" line is then broken and the instrument connected in series. The circuit is then re-energized, and the current reading taken.

The current transformer may be purchased from an electrical hardware supplier. However, a very acceptable one may be constructed by re-winding a small step-down transformer. In particular, I have used a transformer that had a 115-volt primary winding and a 24-volt 1.3-ampere secondary. Units of this rating may be purchased at a local radio store or perhaps found in the junk box. The case/clamp on this unit was pried open and the laminations removed one at a time and carefully saved.

The laminations are held together only with varnish, and the "I" pieces can easily be split off with a putty knife. The associated "E" section can then be split loose and removed. When the core is completely off, the secondary winding (which is usually on the outside) can be stripped off. If the primary winding is intact, it may be left on the bobbin. For a transformer of about this rating, namely 31 volt-amperes, the windings are usually run at 8 to 9 turns per volt. An eight-turn secondary can then be wound on the bobbin using #10 B&S gauge insulated magnet wire. Be careful to ensure that the secondary is well insulated from the primary and from the core eventually. This can be done with certain types of tape. However, conventional masking tape in several layers is adequate if the transformer is not allowed to run too hot. "Too hot" in this case would be about 130°F or where the unit would be uncomfortable to touch.

At this point the reassembly of the core can begin. It is important to resist the temptation to make all the "E" laminations face the same direction. It does make reassembly easier, but it also makes for a very poor transformer. Also try to make sure that all the laminations get back in. The easiest technique is to put in 3 or 4 "E's" facing in alternate directions and then slip the necessary "I's" between them. When these are all in place, the core cover may be clamped around them and the retaining tabs hammered back down while the unit is clamped in a vise. A little epoxy inside the core cover will not hurt and might help prevent the unit from buzzing.

If these steps have been followed carefully, you should now have a transformer with a 110-volt primary and a 1-volt 30-ampere secondary. The secondary winding will actually withstand considerably higher currents for short intervals. The next step is to connect the original primary to 110 volts and measure the primary and secondary voltages with no-load on the secondary. The primary probably consists of about 880 turns of No. 30 wire. The ratio of the two voltages may be used to find the current transformation ratio:

$$\frac{\text{Voltage Primary}}{\text{Voltage Secondary}} = \frac{\text{Current Secondary}}{\text{Current Primary}} = \frac{\text{Turns Primary}}{\text{Turns Secondary}}$$

Now we come to the purpose of the resistor "R" shown in Fig. 1-2. Let us suppose that we would like to arrange the unit so that 30 amperes flowing through the heavy winding would give us a 30-volt output on the light winding. The current in the light winding should be 30 amp/110 or 0.273 amperes. If we force this to flow through a 110-ohm resistor, there will be a 30-volt drop across the resistor. The resistor will be dissipating a power of:

$$\text{Power} = \text{i}^2 \text{ R or } (0.273)^2 \times 110$$
$$= 8.2 \text{ watts}$$

The 273 milliamperes in the light winding is still within reasonable limits, and the unit should run relatively cool. An ordinary multimeter with ac scales would be adequate to measure the voltage on a 1-volt-ampere basis.

One of the advantages of the current transformer is the fact that it can be used to determine the waveshape and the phase angle of the current being drawn. The waveshape is important since a great many modern devices employ silicon control rectifiers (SCRs) in a phase-control circuit to regulate the current being drawn by a controlled appliance. These units tend to mutilate the waveshape. The mutilated waveshape will not read properly in an ordinary ammeter.

# Power factor correction

We had earlier mentioned the fact that the power companies tend to protect themselves by correcting the power factor of their system. This is done for a variety of reasons, including voltage regulation. However, it is usually done to keep the $\cos \theta$ term close to unity so that they have a kVA/kW ratio close to unity. The system losses rise as the square of the system current and a power factor of unity assures that current in the system is minimal for the actual number of watts being consumed.

To try to bring this point home, I will refer to a real physical example, the motor that drives the drill press in my shop. This is an elderly ball-bearing unit that drives the drill press quill, which runs very freely. The drill press is also equipped with a 100-watt lamp. Figure 1-8 shows the oscillograms of the units with a scale change. It can be seen that both units distort the waveform from a pure sine wave, and that the lamp current leads the voltage slightly and the motor current lags the voltage by quite a bit. Now the motor is absorbing 7.2 amperes peak $\times 0.707 = 5.09$ amperes rms and the line voltage is 163 V peak $\times 0.707 = 115$ V rms for a total of $115 \times 5.09 = 586.62$ volt amperes. However, the actual power being drawn is:

$$115 \text{ V} \times 5.09 \text{ amps} \times \cos 84.5° = 56.23 \text{ watts}$$

The disparity between the volt amperes and the watts is a ratio greater than ten.

Since the power dissipated in the generator winding is proportional to $\text{i}^2$, we see that our generator and wiring is dissipating over 100 times more than is necessary for the work being accomplished.

Now referring to Fig. 1-9, we see that a large condenser (or capacitor) has been added with a switch and three ammeters. From the switch open/closed data we see:

- The addition of the condenser cuts the line current down to exactly the value to supply only the real work.
- The sum of the currents at junction J is not zero as we would expect it to be in the dc case.

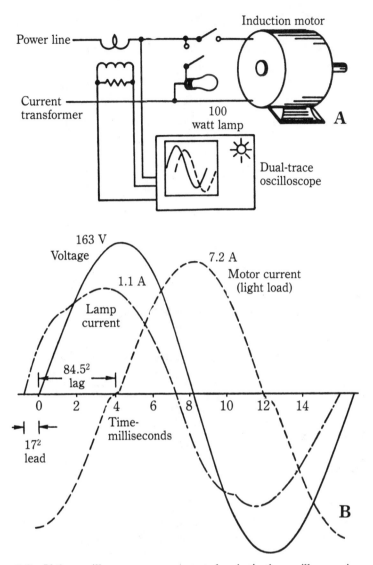

**1-8**  If the oscilloscope were connected as in A, the oscillogram in B would appear.

The advantages of reducing the current in the generating apparatus by this magnitude is relatively obvious. With any given kVA rating of generator we could run more than 10 times as many induction motors if they were carefully tuned out to eliminate the reactive current component.

Program 1-3 shows the mechanism for computing the currents, voltages, and capacitance. If the real part of the current is $i \cdot \cos(th)$, it should come as no surprise that the reactive part is $i \cdot \sin(th)$. We have simply broken the current into two components. One, the real part (or power part), is exactly in phase with the volt-

Capacitor C1 is 117 $\mu$F

With S1 open:

A1 = 5.09 A          A2 = 0          A3 = 5.09 A

With S1 closed:

A1 = 0.488 A          A2 = 5.07 A     A3 = 5.09 A

**1-9**   Determining power factor correction.

### Program 1-3. Real and reactive, by Jack Kuecken

```
100   REM ********************************************
105   REM           REAL AND REACTIVE
110   REM           BY JACK KUECKEN
115   REM               01/16/85
120   REM ********************************************
130   HOME
140 PI = 3.14159265
145   PRINT "USE RMS VALUES FOR VOLTAGE AND CURRENT"
146   PRINT ""
150   INPUT "ENTER MAGNITUDE OF CURRENT IN AMPERES ";I
160   INPUT "ENTER PHASE ANGLE IN DEGREES ";TH
170 TH = TH * PI / 180: REM  CONVERT TO RADIANS
180   IF TH > 90 OR TH <  - 90 THEN  PRINT "ANGLE MUST BE LESS T
      HAN +- 90 ": GOTO 160
190   INPUT "ENTER VOLTAGE ";V
200   PRINT "REAL POWER IS ";(V * I *  COS (TH));" WATTS"
210   PRINT "REACTIVE V.A.= ";(V * I *  SIN (TH))
220   PRINT "REAL CURRENT IS ";(I *  COS (TH));" AMPS"
230   PRINT "REACTIVE CURRENT IS ";(I *  SIN (TH));" AMPERES"
240   IF TH < 0 THEN  PRINT "INDUCTIVE CORRECTION REQUIRED": STOP

250   PRINT "CAPACITANCE REQUIRED FOR P.F. CORRECTION IS"
260   PRINT ((I *  SIN (TH)) * 1E6) / (377 * V);" MICROFARADS"
270   PRINT ""
2000  STOP
```

age, and the other (or reactive part) is +− 90 degrees out of phase. The power factor correction simply requires that we add another reactive component of equal magnitude but opposite sign to cancel the existing reactive part. The impedance for a capacitor is 1/(W·C) where W is the radian frequency, or 2·PI·F, which numerically works out to 366 for F = 60 Hz. The required impedance is (i·sin(th))/ V. We have multiplied by 1E6 to get the result in microfarads, which is the most commonly used unit. It should be noted that the capacitors must be units rated for ac duty.

In subsequent chapters we shall investigate system control in more detail. The next topic on the agenda is a brief treatment of the financial aspects of home electrification. It is very rare that one does anything where "money ain't no object," and home electrification is probably no exception.

# 2
# Economics

ON THE TYPICAL COLLEGE CAMPUS, ONE OF THE TIME-HONORED AND popular arguments concerns the question of the difference between an engineer and a scientist. Just what makes up this difference? It is not necessarily a matter of academic degrees since a great many working engineers have their Ph.D. One widely told story begins with the psychology department that ran an experiment to determine the differences in outlook. They invited an engineer, a physicist, and a mathematician in, ostensibly for an interview. The engineer had the first appointment. When he arrived and walked into the room, he saw the draperies in flames and a bucket of water in the center of the room. He threw the water on the fire and extinguished it. The physicist, on finding the same situation, made some mental calculations, poured some of the water into his hat, and threw that on the flames. He managed to extinguish the fire while conserving some of the water. The mathematician's response was puzzling. He arrived and saw the fire, did some mental calculations, and then turned and left, leaving the drapes burning. When asked about his behavior, he said, "I knew how to solve the problem!"

The moral is that the engineer took the direct action that solved the problem, and he did not waste his time trying to conserve a resource when that effort was not economically justified.

Engineering is supposed to be a practical art and as such should be concerned with arriving at the economically viable solution. This chapter provides some economic considerations involved in generating home electricity.

As a matter of course, most engineering projects begin with a mathematical financial model. Electric power companies are particularly concerned with economics and pursue the analysis of their business in great detail. Let us briefly examine a scenario concerning a medium-sized electric utility to see the things that they consider.

As a general matter, life holds few surprises for an electrical utility company in terms of business growth. In general, the wattage figures for previous years can be smoothly extrapolated and the engineering staff will be able to predict quite accurately the date when a new piece of machinery will be required on line. Eventually, it is predictable when the existing generating plants can no longer be practically enlarged and a whole new facility will be required. The energy crunch a few years back stimulated energy consciousness and changed the slope on the growth curve, but the slope remains.

From 1884 to the mid-1970s the electrical utilities grew at a rate of 8% per annum, doubling every 9 years. Since 1974 the growth has slowed to 2.5%, doubling every 28 years.

The point where the existing facilities cannot be expanded to meet the requirement for power is a significant one. The time required to build even a new thermal plant has expanded to 12.5 years and involves environmental impact statements, construction approvals, etc., etc. If the utility is going to decide to construct a new facility, it must make the decision in a timely fashion in order to have the facility ready to go when needed.

The utility usually has the option of purchasing power from the regional power pool, and as a matter of fact, most do so on a daily basis. Whenever the load pushes up to 85% of spinning capacity, a dispatcher obtains quotes for additional power from the pool members. He then decides whether it is cheaper to buy the required power (if available) or to start up another turbine (if he has one). Conversely, as his load drops below 85% he has the option of selling power or, at some stage, shutting down a turbine. He will know the operating cost of each of his generators to the thousandth of a penny per kWh and he will continually jockey to minimize his cost. The purchased power might be very expensive, coming from a gas turbine, or very cheap, coming from an old hydro plant that is fully amortized (however, most companies keep their hydro power for themselves). In terms of cost, hydro is generally cheapest, nuclear is next, then come coal-fired steam, oil-fired steam, and lastly gas turbine. Hydro becomes cheap after many years when the equipment is paid off or amortized. A number of hydro plants have operated 50 or 75 years with only modest cost for upkeep and modernization. Unfortunately, there are not enough good hydro sites to supply more than a small fraction of the nation's power requirements. At the other end of the scale, the gas turbine runs on expensive fuel, but it is compact and relatively cheap per installed kW. Probably the most significant reason why gas turbine installations exist at all is that they are nearly instant starting. This is a property shared with hydro power, which can be put on line nearly as fast as the valves can be opened. All the steam plants, including nuclear, take hours to bring on line from a cold start. The dispatcher will generally run hydro near capacity and nuclear near capacity if possible. Unexpected peaks can be handled with gas turbine power. If necessary, the gas turbine can be used while a coal- or oil-fired steam plant is brought up. The slowness of response from these systems is the reason for the use of the term "spinning capacity," which refers to the capacity of the machines to start running at any instant and thus be available to handle load peaks. The utilities are required to operate at no more than 85% of their spinning capacity to allow for sudden load increases.

# Return on investment

Just as important as the direct operating cost is the matter of return on investment. As a matter of fact, the return might well be the major part of the real operating cost of the system. When the utility decides that a new plant is required, they may elect one of several routes to finance the new plant. If they have the money in cash reserves, they may simply elect to pay for it out of these reserves. However, if they do not have the money in cash reserves, they might elect to borrow the money. The borrowing can take one of several forms. It may be a direct mortgage such as one obtains on his home or it may be debt obligation in the form of bonds. One additional form of financing comes from the sale of common stock. Regardless of how it is done, there is always a cost associated with the financing. In the case of the mortgage, one obviously must pay interest and principal. In the case of the bonds there are generally coupon payments, and then the face value of the bond must be returned at some time. If the money is raised by a stock issue, there is no promise of payback; however, there is a transfer of equity to the purchaser and an implied promise of dividends. Most utilities stocks are purchased because of large and reliable dividend payments. Even in the case where construction can be financed from cash reserves, there is a cost involved since the cash reserves are almost certainly invested somewhere and removal of the funds will cost a loss in interest.

This is not just an accountant's gimmick, it represents a very real cost, which is chargeable against the operation of the plant. Let us take a small concrete example to make the point a little more strongly. Suppose there was an energy-saving gadget available for $100 and the gadget had a life of 7 years. Further, let us suppose that the money currently resides in a tax-free municipal bond fund that pays 10% per annum. I am reinvesting the dividends so that the interest is, in effect, compounded.

The value of the account is:

| End of Year | Value |
|:-----------:|:------:|
| 1 | $110 |
| 2 | 121 |
| 3 | 133.10 |
| 4 | 146.41 |
| 5 | 161.05 |
| 6 | 177.16 |
| 7 | 194.87 |

Over the seven years, if the gadget has not saved me close to $200, I would have been better off simply to leave the money in the bond fund.

There are several significant points about this. Note that the expression for compound interest is nonlinear, that is, at the end of the first year I collected $10 in interest, whereas at the end of year 7, I collected $17.72, and at the end of year 8, I would be collecting $19.49. This tells us that on a long-payback item, the payback necessary to recover the investment grows very rapidly.

It also should be noted that the payback came back to us in the form of a series of monthly payments rather than as a lump sum at the end of the seven years; consequently, we had the use of at least part of the money for some years. Properly, therefore, we should credit the payback with interest; provided we put it into the bank as it was received, the interest has a compounding effect, and the account operates as an annuity.

# The annuity

Suppose that every month we faithfully take the savings accrued from our gadget and faithfully deposit it in the bank in a day-of-deposit-day-of-withdrawal account. Once a month the bank would calculate interest on the basis of the money in our account and add that and our monthly deposit to arrive at our new balance. At the end of a period like 7 years we would be collecting a significant amount of income from interest.

It seems to me that in this day and age, one of the simplest ways of explaining a concept such as this is through a simple computer program. Program 2-1 is written in Applesoft, a simple dialect of basic. Lines 160 through 190 compute the interest for a given month based upon the balance, and then add that interest and the new payment. After 12 iterations the program prints the results for the year and cleans out the interest summer (I2) and starts over.

**Program 2-1. Annuity value, by Jack Kuecken**

```
100   REM ::::::::::::::::::::::::::::::::::::::::::::::::::
105   REM            ANNUITY VALUE
110   REM          BY JACK KUECKEN
115   REM             02/21/1984
120   REM ::::::::::::::::::::::::::::::::::::::::::::::::::
125   HOME
130   PRINT "ENTER PAYMENT PER MONTH $"
135   INPUT P2
140   PRINT "ENTER INTEREST RATE IN %/YR."
145   INPUT R
150 R2 = R / 1200
151   REM  COMPUTE INTEREST RATE PER MONTH
152   PR# 1: HOME
155   PRINT "FOR AN ANNUITY WITH $ ";P2;" INPUT PER MONTH AT ";R
      " % INTEREST"
157   PRINT ""
160   FOR K = 1 TO 12 ·
170 I = P3 * R2
172   REM  COMPUTE INTEREST FOR THE MONTH
175 I2 = I2 + I
180 P3 = P3 + I + P2
185   REM  COMPUTE NEW BALANCE ADDING INTEREST AND MONTHLY PAY'T
      ·
190   NEXT
200 N = N + 1
210   PRINT "AT THE END OF YEAR ";N;" THE PRINCIPAL IS $ ";P3
220   PRINT "AND INTEREST FOR THE YEAR WAS $ ";I2
230 I2 = 0
240   PRINT ""
250   GOTO 160
```

From the runout on the demonstration at the bottom of Table 2-1 we see that a savings of $1.87 per month at a 6% rate would be adequate to put our bank deposit back in the shape it would have been in if we had not taken out the $110 for the gadget. This is considerably less than $194.87/(7·12) = $2.32/month.

The lower interest rate was selected to represent a DOD/DOW account, which pays less than a long-term time deposit in the tax-free municipal fund.

### Table 2-1. For an annuity with $1.87 input per month at 6% interest

At the end of Year 1 the principal is $23.0675016 and interest for the year was $.627501637

At the end of Year 2 the principal is $47.5577563 and interest for the year was $2.05025466

At the end of Year 3 the principal is $73.5585164 and interest for the year was $3.56075999

At the end of Year 4 the principal is $101.162946 and interest for the year was $5.16442997

At the end of Year 5 the principal is $130.469957 and interest for the year was $6.86701082

At the end of Year 6 the principal is $161.58456 and interest for the year was $8.67460312

At the end of Year 7 the principal is $194.618244 and interest for the year was $10.5936838

At the end of Year 8 the principal is $229.689373 and interest for the year was $12.6311291

# Taxes

At the present writing, taxes enter into the equation from two standpoints. In general, the interest income is taxed as ordinary income and the interest paid on a loan is deductible from your income. In both cases you are interested in your incremental tax rate or the tax rate paid on the last dollar of your income. This can be found by looking at the tax tables that went with your last tax return. For both federal and state taxes there is usually a table saying that if your income was between $xxxx and $yyyy your tax is $zzz plus nn% of everything over $xxxx. The nn% is the incremental tax rate. If your income had been a dollar higher or dollar lower, your tax would have changed by that many cents. In the main IRS tax tables, the taxes are broken down every $50. For these one can take the adjacent readings and calculate the percentage. For example, suppose that your taxable income (line 37 on your 1040) was $37,953. The table might have an entry for:

$37,900 to 37,950   tax is $7,173
$37,950 to 38,000   tax is $7,190

In effect, on the last $50 a tax of $7,190-$7,173 or $17 was charged, or roughly 34% incremental tax.

There is generally an exclusion allowed on interest and dividend income; permitting exclusion of perhaps the first $400 or so. These items tend to change from year to year, so you really should consult the instructions for preparing the form 1040 as well as the instructions for your state income tax (if you have one).

Obviously, this form of anlysis is an attempt to predict the future. We are doing it with a computer based upon past performances and mathematical extrapolations of known past performances of things. This is a little more dignified than attempting to do the same thing with tarot cards, knuckle bones, or the entrails of a goat; however, it remains an excercise in fortune-telling. A major tax simplification could remove the tax deductions and perhaps provide a fixed marginal tax; however, in the meantime we shall continue to extrapolate since this is the best we know how to do.

# Inflation

Another factor that is probably worth including is inflation. Throughout most of recorded history, the price of things in general has tended to grow with the passage of years. There are a few exceptions, such as integrated circuits, whose prices have tended sharply down with time (The Learning Curve); however, over the same time span most other prices have risen. In general, electric energy prices have increased somewhat slower than inflation across the board; however, it is fair to assume that electrical energy will cost more in the future than it does today. Program 2-1 can be adjusted for inflation with the addition of one line:

245 P2 = P2•1.03:rem 3% adjustment for inflation

The data might be a little clearer also if we add:

205 print "for monthly payment $";P2

This helps to keep track of what the payment is at any point in time. Note that the inflation adjustment is made only annually since the power companies will seldom be allowed to adjust their rates even that often. Table 2-2 shows that an inflation rate of 3% yields results differing significantly from the uninflated value for our investment payback. Of course, if double-digit inflation were to return, the payback on our investment would really improve.

**Table 2-2. For an annuity
with $1.87 input per month at 6% interest
(Inflation at 3%)**

For monthly payment $1.87
At the end of Year 1 the principal is $23.0675016
and interest for the year was $.627501637

For monthly payment $1.9261
At the end of Year 2 the principal is $48.2497813
and interest for the year was $2.06907971

For monthly payment $1.983883
At the end of Year 3 the principal is $75.6980348
and interest for the year was $3.64165742

For monthly payment $2.04339949
At the end of Year 4 the principal is $105.573406
and interest for the year was $5.35457713

For monthly payment $2.10470148
At the end of Year 5 the principal is $138.047619
and interest for the year was $7.21779528

For monthly payment $2.16784252
At the end of Year 6 the principal is $173.30365
and interest for the year was $9.24192144

For monthly payment $2.2328778
At the end of Year 7 the principal is $211.536444
and interest for the year was $11.4382597

For monthly payment $2.29986413
At the end of Year 8 the principal is $252.953666
and interest for the year was $13.8188528

For monthly payment $2.36886005
At the end of Year 9 the principal is $297.776515
and interest for the year was $16.3965289

For monthly payment $2.43992585
At the end of Year 10 the principal is $346.240577
and interest for the year was $19.1849512

For monthly payment $2.51312363
At the end of Year 11 the principal is $398.596731
and interest for the year was $22.1986709

# The mortgage

For a significant-sized investment, you might have to borrow the money. You and I generally cannot issue bonds or sell stock shares; therefore, the borrowing generally takes the form of a loan. There are many types of loans and many ways of repaying them; however, I will presume here that the loan is similar to a home mortgage where a payment of fixed size is made every month over a period of years established at the time the loan was issued. Interest is computed on the unpaid balance and the fixed payment is made up of one month's worth of interest and some payment on the principal. At the start of the loan period the interest is large, so the fixed payment minus interest leaves little for reduction of principal. Only well into the life of the loan does the reduction of principal in the monthly payment become larger than the interest charge.

The assumption is made that the fraction of interest equal to your marginal tax rate will appear as a tax deduction. A word is in order here about the difference between a TAX CREDIT and a TAX DEDUCTION. Suppose that you are earning at a rate such that, after all other deductions, your taxable income is $37,953, on which the tax (married, joint) is $7,190. Now you note that you paid $2,000 in interest on your house last year. This is a TAX DEDUCTION, which means that you figure your income tax on $2,000 less, or $35,953, on which sum the tax is, say, $6,530—a savings of $660 or 33% of the $2,000 (remember the marginal tax rate?).

A TAX CREDIT of $2,000 would have been subtracted from the $7,190, directly reducing your taxes to $5,190. Only rarely is one entitled to a tax credit.

One of the points to be made here is the fact that "It's deductible" definitely does not mean "It's free." In the example above you still were out 67% of the $2,000.

If we return to the mortgage, the calculation of the monthly payment that will satisfy the loan in the specified period is not immediately obvious. The amount of principal and interest in each monthly payment each varies for every payment in the life of the loan. The general case solution can be developed by setting up and solving a differential equation. The detailed math is outside the scope of this book; however, the solution is presented in the computer program that follows.

Program 2-2 presents a mortgage evaluation calculation that can be used to determine the approximate cost of a loan. The results turn out within a few cents of the tabular solutions used by the bank in setting up the loan. The computation for the monthly payment, designated K1, is performed on lines 340 and 350. The notation EXP( ) means the natural log base "e" raised to the power. The R1/12 term is, of course, the interest rate per month. The business on lines 360, 370, and 380 is put there to round off the monthly payment to even cents. The Apple has no single command that does this. On a long-term loan the fractional cents would incur a significant error, and of course it is impossible to make monthly payments in fractional cents.

### Program 2-2. Mortgage loan cost analysis, by Jack Kuecken

```
100   REM   ************************************************
105   REM       MORTGAGE LOAN COST ANALYSIS
115   REM       BY JACK KUECKEN
120   REM           08/08/84
125   REM   ************************************************
130   HOME
150   INPUT "ENTER MORTGAQGE INTEREST RATE % ";R
155 R1 = R / 100
160   PRINT "INPUT BANK LOAN PERIOD"
165   INPUT T1
190   PRINT "INPUT AMOUNT OF LOAN"
200   INPUT L1
205 L = L1
210   INPUT "ENTER YOUR INCREMENTAL TAX RATE % ";TR
220 TR = TR / 100
300   REM ********************************************
310   REM   THE CALCULATION
320   REM    FOR THE PAYMENT SIZE
330   REM $$$$$$$$$$$$$$$$$$$$$$$$$$$$$$$$$$$$$$$$$$$$$$
340 A1 =   - (R1 * T1)
350 K1 = ( - L1 * R1 / 12) / ( EXP (A1) - 1)
```

```
360  K1 = (K1 * 100) + .5
370  K1 =   INT (K1)
380  K1 = K1 / 100
500     REM   $$$$$$$$$$$$$$$$$$$$$$$$$$$$$$$$$$$$$$$$$$$$$
510     REM        THE DATA OUTPUT
520     PR# 1
530     PRINT "FOR A TOTAL MORTGAGE OF $ ";L;" ON A PERIOD OF ";T1
        ;" YEARS"
540     PRINT "WITH INTEREST = ";R1
550     PRINT "THE FIXED PAYMENT IS ";K1;" PER MONTH"
595     PRINT ""
610     REM ****************************************
620     REM        CALCULATE PRINCIPAL AND INTEREST
625  J = 1
630     FOR I = 1 TO 12
640  IN = L * R1 / 12
642  IN =   INT ((IN * 100) + .5):IN = IN / 100
650  IY = IY + IN:IT = IT + IN:FI = FI + (TR * IN)
652  PY = K1 - IN
654     REM  SUBTRACT INTEREST FROM PAYMENT
656  L = L - PY
658  L =   INT ((L * 100) + .5)
660  L = L / 100
670     NEXT I
680     PRINT "AT THE END OF YEAR ";J;" THEN REMAINING PRINCIPAL I
        S "
690     PRINT "$ ";L;" AND INTEREST PAID IS $ ";IY
692     PRINT "YOUR TAX DEDUCTION FOR INTEREST IS $ "(IY * TR)
695     PRINT ""
700  IY = 0
710  J = J + 1
715     IF J > T1 GOTO 750
720     IF L > K1 GOTO 630
730     REM ***************************************************
740     REM        THE SUMMARY STATEMENT
750  PT = K1 * 12 * T1 + L
760     PRINT
770     PRINT "YOUR PAYMENTS TOTALED $ ";PT;" FROM WHICH YOU DEDUC
        TED $ ";FI" IN TAX REFUNDS"
790     PRINT "INTEREST COST AFTER TAXES WAS $ ";(IT - FI)
2990    PR# 0
3000    STOP
```

The routine from 610 through 720 calculates the principal and interest on a running month-by-month basis in a bookkeeping mode. It prints a summary for every year. At the conclusion of the loan the routine at 740 sums up the total payments, including the small error caused by the rounding, and presents total interest and deductions over the life of the loan.

The printout of Table 2-3 shows that we would have to pay out nearly half again the price of the gadget; however, some of the interest cost would be recovered at tax time. Note that it takes about 4.4 of the 7 years to reduce the loan principal to half the initial amount.

**Table 2-3. For a total mortgage of $100 on a period of
7 years with interest = .12 the fixed payment is 1.76 per month**

---

**For a total mortgage of $100 on a period of 7 years
with interest = .12
the fixed payment is 1.76 per month**

At the end of Year 1 then remaining principal is $90.36
and interest paid is $11.48
Your tax deduction for interest is $3.7884

At the end of Year 2 then remaining principal is $79.48
and interest paid is $10.24
Your tax deduction for interest is $3.3792

At the end of Year 3 then remaining principal is $67.26
and interest paid is $8.9
Your tax deduction for interest is $2.937

At the end of Year 4 then remaining principal is $53.47
and interest paid is $7.33
Your tax deduction for interest is $2.4189

At the end of Year 5 then remaining principal is $37.92
and interest paid is $5.57
Your tax deduction for interest is $1.8381

At the end of Year 6 then remaining principal is $20.4
and interest paid is $3.6
Your tax deduction for interest is $1.188

At the end of Year 7 then remaining principal is $.65
and interest paid is $1.37
Your tax deduction for interest is $.4521

Your payments totaled $148.49 from which you deducted $16.0017 in tax refunds
Interest cost after taxes was $32.4883001

---

For Table 2-4 the amount has been expanded to $1000. The parameters were
explored from two standpoints: at a fixed 20-year period with the interest rate as
the independent variable, and at a fixed 12% interest rate with the period as the
independent variable. Note that the payment climbs rapidly as function of interest
rate, and the total cost does also. With time as an independent variable however, it
can be seen that payment size decreases slowly beyond 15 years. On the other
hand, the total cost rises rapidly beyond 15 years.

It was found that the half-pay point on the 20-year loan is a rather slow function of interest rate ranging from 14.5 years at 10% to 16 years at 16% interest.

The main lesson is that a great deal of money can be saved by paying a little
more in monthly payments on a shorter-term loan.

# Backing in

Often it is easy to see how we can save a certain amount per month, and then the
question becomes "How much can I practically spend to achieve those savings?"

### Table 2-4. For a $1000 Mortgage
### (20-year period)

| Interest % | Payment/month $ | Total cost |
|---|---|---|
| 10 | 9.64 | 2,321.20 |
| 12 | 11.00 | 2,650.91 |
| 14 | 12.42 | 2,999.85 |
| 16 | 13.90 | 3,357.58 |

### For 12% interest and variable period

| Period | Payment/month | Total cost |
|---|---|---|
| 5 yr | 22.16 | 1,336.44 |
| 10 | 14.31 | 1,725.75 |
| 15 | 11.98 | 2,167.09 |
| 20 | 11.00 | 2,650.91 |
| 25 | 10.52 | 3,179.24 |
| 30 | 10.28 | 3,721.99 |

The simple Program 2-3 is designed to answer this question. It will work with periods including decimal fractions of years. And it is simple enough to do on a pocket calculator if it has logs and antilogs.

The first thing to do is to determine how much money you can expect to save over a period of years. This is done with the annuity program with the result

### Program 2-3. Allowable expense, by Jack Kuecken

```
100    REM **********************************************
110    REM          ALLOWABLE EXPENSE
115    REM          BY JACK KUECKEN
120    REM             12/28/85
125    REM **********************************************
150    HOME
160    INPUT "ENTER AMOUNT IN $ ";S
170    INPUT "ENTER PERIOD IN YEARS";T1
180    INPUT "ENTER TAX ADJ. INTEREST RATE % ";R
190 R1 = R / 100
200 R2 = 1 + R1: REM   THE INTEREST MULTIPLIER
210 R3 =  LOG (R2)
220 R4 = T1 * R3: REM   RAISE IT TO THE POWER
230 R5 =  EXP (R4): REM   BACK TO THE GROWTH RATIO
250 P1 = S / R5: REM   CALCULATE THE AMOUNT
252    PR# 1
260    PRINT "$ ";P1;" INVESTED AT ";R;"%"
270    PRINT "FOR ";T1;" YEARS PRODUCES ";S
280    PR# 0
290    GOTO 100
1000   STOP
```

### Table 2-5. For an annuity with a
### $55 input per month at 3.6% interest

For monthly payment $55
At the end of Year 1 the principal is $670.999639
and interest for the year was $10.9996386

For monthly payment $56.65
At the end of Year 2 the principal is $1386.68784
and interest for the year was $35.8882013

For monthly payment $58.3495
At the end of Year 3 the principal is $2149.30411
and interest for the year was $62.4222643

For monthly payment $60.099985
At the end of Year 4 the principal is $2961.18802
and interest for the year was $90.6840901

For monthly payment $61.9029846
At the end of Year 5 the principal is $3824.78345
and interest for the year was $120.759613

For monthly payment $63.7600741
At the end of Year 6 the principal is $4742.64294
and interest for the year was $152.738595

For monthly payment $65.6728763
At the end of Year 7 the principal is $5717.43223
and interest for the year was $186.714782

For monthly payment $67.6430626
At the end of Year 8 the principal is $6751.93506
and interest for the year was $222.786075

For monthly payment $69.6723545
At the end of Year 9 the principal is $7849.05801
and interest for the year was $261.054703

For monthly payment $71.7625251
At the end of Year 10 the principal is $9011.8357
and interest for the year was $301.627402

For monthly payment $73.9154009
At the end of Year 11 the principal is $10243.4361
and interest for the year was $344.615604

For monthly payment $76.1328629
At the end of Year 12 the principal is $11547.1661
and interest for the year was $390.135632

For monthly payment $78.4168488
At the end of Year 13 the principal is $12926.4772
and interest for the year was $438.308906

For monthly payment $80.7693543
At the end of Year 14 the principal is $14384.9716
and interest for the year was $489.26215

For monthly payment $83.1924349
At the end of Year 15 the principal is $15926.4084
and interest for the year was $543.127617

shown in Table 2-5. For the table we assumed an inflation rate of 3% in the price of electric power and a tax de-rated interest rate of 3.6%. The amounts work out to fairly tidy sums over the years. Our problem now relates to finding out how much money, invested at compound interest would be required to earn these amounts.

In computing compound interest, as with our table for the $110, if you stop and think about it, what one does is to multiply the principal by the growth rate as many times as there are years in the period. For example, if the interest rate is 4%, the growth rate is 1.04/year.

Then, if we have a principal "P" and a period of three years, the calculation will be:

$$\text{FINAL BALANCE} = P \cdot 1.04 \cdot 1.04 \cdot 1.04$$

which could also be written:

$$\text{FINAL BALANCE} = P \cdot (1.04)^3$$

In others words, the growth rate is raised to the power of the number of growth periods. Now, it is easy to raise a number to some whole number power, but how does one raise a number to fractional power? The answer is through the use of logarithms:

$$\text{Log}(R)^T = T \cdot \text{Log}(R)$$

All we have to do is take the log of the rate "R" and multiply it by the number of periods "T" and then take the antilog of the product to obtain the growth ratio or the ratio between the initial principal and the final balance. Program 2-3 does this on lines 210 through 230. The sum from the annuity table is then divided by the growth ratio to determine how much would have to have been invested to obtain that final balance. The runout of Table 2-6 shows the results at 5, 10, and 15 years. If the investment in the plant is less than that, then you are ahead of the game. Note that the assumption is that there is no cost after the initial investment. You have to make some judgment about how long the system would operate without your having to pour more money into it. Obviously, if it goes 10 years rather than 5 without investment, one could afford to spend more money on the installation in the first place, and this fact is shown in the table. If the installation is expected to have some regular cost, for example oil and maintainance billed on a regular basis by the dealer, this could be accounted for by subtracting it from the initial $55 of savings used in the annuity program.

**Table 2-6.**

---

$2957.17535 invested at 5.28%
for 5 years produces 3824.78

$5387.0959 invested at 5.28%
for 10 years produces 9011.84

$7360.88009 invested at 5.28%
for 15 years produces 15926.41

---

# Investment payback period

A criterion frequently applied to any investment scheme is "How long will it take me to get my money back ?" After all, a scheme that will pay your money back in 55 years might not be of much interest if you are now 60. Program 2-4 combines some of the functions of previous programs to determine the payback period on an investment. It accounts for the growth of the investment (if it has been left in the bank and also accounts for your incremental tax rate). It accepts the net savings per month and lets these grow with an assigned inflation rate. Line 260 computes the interest minus tax growth on the investment and line 270 computes the inflated savings per month. Obviously, if the savings are less than the interest growth of the investment, you will never get your money back. Provided that the monthly savings are greater than the growth of the investment, the program proceeds to 300 and beyond. Line 385 prints to the screen the current values of principal and monthly savings and counts the months and years. If the count of years gets to 30 with no payoff, the program throws in the towel on line 425, since it is silly to carry this form of projection so far into the future. However, if the payoff is achieved in less than 30 years, the payoff time is printed in lines 500 and 510. This is not a fast-running program; however, watching the numbers go by gives one a feel for the interaction of the factors. Table 2-7 shows the last readouts from an example.

### Program 2-4. Investment payback period, by Jack Kuecken

```
100    REM  ***********************************************
105    REM           INVESTMENT PAYBACK PERIOD
110    REM              BY JACK KUECKEN
115    REM                 07/15/85
120    REM  ***********************************************
130    HOME
140    PRINT "ALL PERCENTAGES ARE PER YEAR"
150    INPUT "ENTER AMOUNT OF INVESTMENT $";IV
160    INPUT "ENTER INTEREST RATE % ";R1
170    INPUT "ENTER YOUR INCREMENTAL TAX RATE % ";R2
180 R1 = R1 / 100:R2 = R2 / 100: REM   CONVERT TO DECIMAL
190    INPUT "ENTER INCOME/REVENUE $";IR
200    INPUT "PER MONTH OR YEAR ? M/Y ";T$
210    IF T$ = "M" GOTO 250
220    IF T$ = "Y" THEN IR = IR / 12: GOTO 250
230    PRINT "IT MUST BE M OR Y"
240    GOTO 190
250    INPUT "ENTER INFLATION RATE % ";R3
255 R3 = R3 / 100
260 II = (1 - R2) * IV * R1 / 12
270 IJ = IR * (1 + (R3 / 12))
280    IF II < IJ GOTO 350
290    PRINT "INVESTMENT CANNOT BE RECOVERED WITH THOSE PARAMETER
       S
300    REM  IK MONTHLY INTEREST GROWTH ON INVESTMENT
310    REM  IL IS MONTHLY INFLATION GROWTH ON INCOME/REVENUE
320    REM  ***********************************************
```

```
330  REM        THE PAYBACK LOOP CALCULATION
350  IK = IV * R1 / 12
355  IM = IM + IK
360  IL = IR * (R3 / 12)
370  IV = IV + IK
375  IR = IR + IL
380  IN = IN + IR
385   PRINT IV,IR
390  N = N + 1
400   IF IN > IV GOTO 499
410   IF N < 12 GOTO 350
420  N = 0:M = M + 1: REM  N IS MONTHS AND M IS YEARS
425   IF M > 30 THEN  PRINT "RECOVERY NOT FEASIBLE ": STOP
430  TT = IM * R3: REM   TAXES
440  IL = IL - TT: REM  PAY THE TAXES!
450   GOTO 350
490  REM *******************************************
495  REM        PAYOFF PRINTOUT
499  PRINT ""
500  PRINT "THE ACTIVITY PAYS OFF IN "
510  PRINT M;" YEARS AND ";N;" MONTHS
2000  STOP
```

### Table 2-7.

| Investment | $1000 |
| --- | --- |
| Interest | 12% |
| Incr. tax rate | 33% |
| Inflation rate | 3% |
| Savings/month | $9 |

| | | | |
| --- | --- | --- | --- |
| 1003.70667 |  | 9.87107878 | |
| 1003.80704 | 1006.92351 | 9.89575647 | 10.6921445 |
| 1003.90742 | 1007.0242 | 9.92049586 | 10.7188748 |
| 1004.00781 | 1007.12491 | 9.94529711 | 10.745672 |
| 1004.10821 | 1007.22562 | 9.97016035 | 10.7725362 |
| 1004.20862 | 1007.32634 | 9.99508575 | 10.7994675 |
| 1004.30904 | 1007.42707 | 10.0200735 | 10.8264662 |
| 1004.40947 | 1007.52782 | 10.0451236 | 10.8535324 |
| 1004.50991 | 1007.62857 | 10.0702365 | 10.8806662 |
| 1004.61036 | 1007.72933 | 10.095412 | 10.9078679 |
| 1004.71083 | 1007.83011 | 10.1206506 | 10.9351375 |
| 1004.8113 | 1007.93089 | 10.1459522 | 10.9624754 |
| 1004.91178 | 1008.03168 | 10.1713171 | 10.9898816 |
| 1005.01227 | 1008.13249 | 10.1967454 | 11.0173563 |
| 1005.11277 | 1008.2333 | 10.2222372 | 11.0448997 |
| 1005.21328 | 1008.33412 | 10.2477928 | 11.0725119 |
| 1005.3138 | 1008.43496 | 10.2734123 | 11.1001932 |
| 1005.41433 | 1008.5358 | 10.2990958 | 11.1279437 |
| 1005.51488 | 1008.63665 | 10.3248436 | 11.1557636 |
| 1005.61543 | 1008.73752 | 10.3506557 | 11.183653 |
| 1005.71599 | 1008.83839 | 10.3765323 | 11.2116121 |
| 1005.81656 | 1008.93927 | 10.4024737 | 11.2396411 |
| 1005.91714 | 1009.04017 | 10.4284798 | 11.2677402 |

**Table 2-7. Continued.**

| | | |
|---|---|---|
| Investment | $1000 | |
| Interest | 12% | |
| Incr. tax rate | 33% | |
| Inflation rate | 3% | |
| Savings/month | $9 | |

| | | | |
|---|---|---|---|
| 1006.01773 | 1009.14107 | 10.454551 | 11.2959096 |
| 1006.11834* | 1009.24199 | 10.4806874 | 11.3241493 |
| 1006.21895 | 1009.34291 | 10.5068891 | 11.3524597 |
| 1006.31957 | 1009.44384 | 10.5331564 | 11.3808409 |
| 1006.4202 | 1009.54479 | 10.5594893** | 11.409293 |
| 1006.52084 | 1009.64574 | 10.585888 | 11.4378162 |
| 1006.6215 | 1009.74671 | 10.6123527 | 11.4664107 |
| 1006.72216 | 1009.84768 | 10.6388836 | 11.4950768 |
| 1006.82283 | 1009.94867 | 10.6654808 | 11.5238145 |

The activity pays off in 8 years and 3 months

*Current val. principal

**Current savings 1 month

Unless you particularly like to hear your printer run, it is usually not worthwhile to have this program print hardcopy. It will run noticeably faster with line 385 deleted and the monthly print to the screen eliminated. Alternatively, you might want to print IV and IN showing current investment and current total savings. One of the uses of this program is to investigate the effects of the inflation rate, the interest rate, and the net savings on the payback period.

# Summary

In this chapter we have attempted to provide you with at least a feeling for the economic aspects of some of the energy-saving schemes. There are, of course, a number of reasons why one would want to generate his own electricity besides economic ones. However, it is probably better to know from the outset if a particular scheme is a financial loser. If the financial analysis indicates that there is little possibility of recovering your investment you might still want to proceed with the effort as a hobby, but at least you are not fooling yourself.

# 3

# Motors and generators

THE BASIC PRINCIPLE UPON WHICH NEARLY ALL ELECTRIC MOTORS AND generators operate is the discovery by Ampere that a wire carrying an electric current experiences a physical force when it is immersed in a magnetic field and the discovery by Oersted that a wire moved through a magnetic field develops a voltage. These actions are complementary; that is, if the moving wire is connected to an external circuit so that the induced voltage can force a current flow, it will be found that the wire develops a mechanical force, which tends to resist the motion. In other words, it becomes necessary to do mechanical work to move the wire.

The converse principle also works. If a current is forced into a wire in a magnetic field, the wire will develop a force that tends to cause it to move. If nothing opposed this motion, the wire would accelerate to a speed where the induced voltage would exactly buck out the external voltage source and would bring the current to zero. If on the other hand, something opposed the motion of the wire, such as an external mechanical load or the inertia of the wire, then the wire would accelerate to a speed where it bucked out a portion of the external voltage and only enough current would flow to accomplish the mechanical work being done. This bucking voltage is termed a *counterelectromotive force* (counter emf) and is responsible for the workable nature of the electric motor.

While we are not interested in learning the details of designing electric motors in this text, we should examine in a qualitative way some of these principles so that we can understand some of the phenomena that are encountered in a small electrical system. Figure 3-1 shows the simplest kind of electric motor or generator. A coil of wire having a number of turns is pivoted so that it can rotate between the poles of a magnet. Beneath this is a series of further simplified figures showing a sort of moving picture reduced to only the wires themselves.

First of all, if the coil is rotating at some constant speed $\omega$ we can see that the coil sides A and B will cut the magnetic lines $\beta$ at the fastest rate when the coils

**3-1**  In A, the simplified motor or generator. In B, if the wires of the coil are wound in a simple helix, it is fairly easy to see that the voltages in the two sides and the successive turns add up.

are in the positions designated 1 and 5, and will not be cutting any lines at all when in position 3. Furthermore, wire A will be cutting the lines in the opposite direction at position 5 to when it was at position 1. Therefore, we would expect that the voltages induced in A and B would be reversed. If the field between the magnet poles is uniform, it can be shown that the induced voltage is proportional to the sine of the angle $\alpha$ since this is the relationship that determines the rate at which the magnetic lines are cut. The machine is obviously producing alternating current of a frequency $\omega$ equal to the mechanical rotation speed of the shaft.

If the multiple turns of the coil are wound in the form of a simple helix it is fairly easy to visualize the fact that the voltages will add together. The voltage induced will be in direct proportion to:

- The strength of the magnetic field
- The rate of rotation, $\omega$, and the radius of the rotor
- The number of turns on the coil
- The sine of angle $\alpha$ (which varies with time).

If we wanted to use this machine for either a motor or a generator, it would be necessary to provide some means of connecting terminals A and B to the outside world. If we wanted alternating current operation, we could make the rotary connection to the outside world via slip rings as shown in Fig. 3-2.

Voltage

**3-2** The slip ring for AC connection.

The commutator bars of Fig. 3-3 not only provide the sliding rotary connection but also permit the outside connection to "swap hands" between terminals A and B so that brush N can always be positive and brush M can always be negative; that is, if the insulating bar is arranged to be vertical when the coil is horizontal, and the brushes are arranged roughly horizontal. We shall explain the term "roughly" in short order.

Voltage

**3-3** The commutator (simplified) permits the outside world to change hands with terminals A and B so that a unipolar voltage is presented on the machine terminals.

There are several things of importance to be noted from the above conclusions. First and foremost is the fact that with any given machine, the faster it turns, the higher the output voltage will be. Now the current carrying capacity of the machine is largely determined by the size of the wire with which the machine is wound. Therefore, the faster the machine turns, the higher the available power. In aircraft usage, where weight is at a premium, advantage is taken of this fact. A bipolar alternator for aircraft use is generally rotated at 400 rev/sec or 24,000 rpm in order to get the maximum output from the smallest and lightest machine. A given size aircraft alternator with 400-Hz output will generally produce 400/60 times as much power as a 60-Hz machine of the same size. Unfortunately the machines are a good deal more expensive since they must be capable of very high speeds without bursting from centrifugal force, and the armature iron must be made of 0.004-inch laminations rather than the 0.014-inch used in the 60-Hz machine. This high speed advantage holds for motors and transformers as well. A 400-Hz transformer requires only about 1/7 as much iron as a 60-Hz design with the same VA rating, but again the laminations must be thin to avoid excessive losses.

Four-pole 12,000 rpm or eight-pole 6,000 rpm machines can often be found at very low prices on the surplus market, and the machines are outstanding performers. However, they cannot be used to run most 60-Hz appliances that have either transformers or motors. An exception is the type of motor used in most single-speed hand tools such as saws, and hand drills. These will run successfully on the high-frequency current, as will incandescent lamps. With most radios, TVs, washers, dryers, and such, the unit will not run at all or will severely overheat on the high-frequency current.

If we are to generate 60-Hz current directly with the machine, we are generally restricted to 60 revolutions per second or 3600 rpm in bipolar machines and 1800 rpm in four-pole machines. Unfortunately, this is a notable restriction. You can see that the speed with which the machine cuts the flux is also a function of the radius of the machine, so that one would expect the 60-Hz machine to be pretty sizeable and indeed this is the case. In most cases I doubt that even a very strong man would be able to lift a 5-kVA, 60-Hz alternator off the floor!

At this point you are probably about to ask, "If the high-frequency devices have so much advantage, why don't we convert the electrical system to 400-Hz operation? The answer is as follows:

- Because of the high speeds and high frequencies, motors, transformers, and gear trains are inherently more expensive. Wider usage would bring the price down but not to the level of 60-Hz components.

- At 60 Hz the transmission line effects begin to be troublesome at 100 miles and become very complicated to manage beyond 350 miles. At 400 Hz the limits reduce to 17 miles and 52.5 miles respectively. An upward shift in the frequency would make the power company interties a great deal more expensive.

- Design of power-station size components becomes very difficult for the high frequencies.

As we shall subsequently see, modern electronics offers the possibility of utilizing the best of both worlds.

# Armature reaction

It was noted earlier that in our machine of Fig. 3-1, we would set the brushes "roughly" horizontal. To explain this, let us look at the top of Fig. 3-4. This illustration again shows the rotor winding simplified to just two terminals. It can be

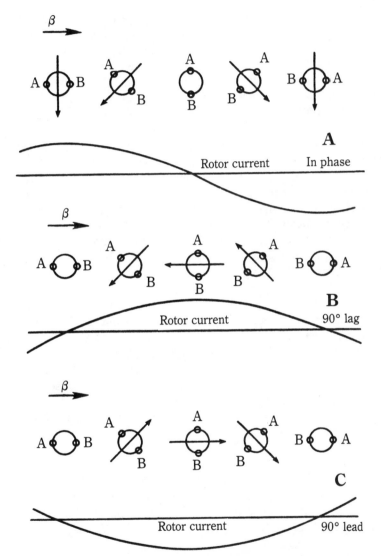

**3-4**  Armature reaction with the rotor winding simplified to just two windings. At A, the machine is connected to an external resistive load. At B, the external load is lagging 90° as if it were a very lightly loaded motor or inductor. At C, the external load is a capacitor so the output current leads the output voltage by 90°.

shown by experiment that a helical winding with a current flowing through it behaves very much like a bar magnet with the magnetic strength being directly proportional to the amount of current. We have also seen that even for the dc machine with the commutator, the voltage developed by the winding is a sine wave with a frequency equal to the rotational speed of the rotor in revolutions per second.

The top illustration shows what happens if the machine is connected to an external resistive load. With this condition, we can see that the current flowing through the rotor tends to generate a magnetic field that fluctuates twice per revolution and is at right angles to the machine flux. The combination of this flux and the machine flux tends to shift the net direction of the flux in the machine so that in the dc machine case the machine produces more output if the brushes are offset to align with the flux field. In modern dc machines this effect is offset by other techniques. However, let us see what this effect does in an alternator.

The next illustration down shows the effect when the external load is lagging 90° as if the external load were a very lightly loaded motor or inductor. It can be seen that the net effect of the armature current is in a direction to directly oppose the magnetic field of the machine. Any reduction in this field will tend to directly reduce the generated voltage. If the field magnetism is not increased the output voltage will sag seriously.

The next figure below shows the effect of the armature reaction when the external load is a capacitor so that the output current leads the output voltage by 90°. Here the action is to strengthen the machine field. This in turn tends to make the output voltage climb and in some cases soar. This is one of the problems associated with energizing a long, unloaded transmission line. The line will act as a large capacitor, and the output voltage of the machine might climb beyond the ability of the field excitation system to control it. Note that this effect is a positive feedback or runaway situation. The higher the output current is, the greater the output voltage is, which in turn increases the output current. The effect is limited by the saturation of the system iron, after which the magnetism increases slowly.

For the inductive load, the armature reaction effect is less disastrous since an increase in current decreases the voltage, which further decreases the current. Unfortunately, the undervoltage can also burn out motors since they might not be able to start and thus might draw excessive current.

This armature reaction is one of the most serious reasons why care should be taken to keep the power factor up in a small system. Unlike the power company, the home electrical system does not have a very large number of independent loads to average out. In some instances, a single motor going into the start cycle might prove to be a major part of the system load.

# A more practical machine

Figure 3-5 shows a somewhat more practical form of ac motor or generator. If winding AB is energized with a direct current, the rotor or armature becomes a two-pole magnet. As it is rotated, it induces a voltage in windings CD and EF. Note

**3-5** A more practical ac motor at A, and the voltages induced in the stationary windings at B.

that the sine wave voltages introduced in these windings are in quadrature, or the voltages are 90° out of phase: voltage CD is zero when voltage EF is maximum, and vice versa. The rotating magnetic field principle was invented by Nicola Tesla and forms the basis for all induction motors.

With dc excitation on the rotor, if the machine were in synchronism with the line and the voltages CD and EF applied from the outside, this machine would run as a synchronous motor and could deliver power through the shaft. Unfortunately, in this form the motor would have no starting torque and would have to be brought up to synchronous speed by some other means.

If winding AB is shorted at terminals A and B and the voltages CD and EF applied from an external source, the device becomes a two-phase induction motor. The quadrature excitation of the two phases produces a rotating magnetic field. This field in turn induces currents in the shorted windings of the rotor, which tend to drag the rotor along with the rotating magnetic field. This type of motor will develop a very sizeable starting torque and will run at some speed lower than the line.

If the rotor is turning at precisely the line speed, say 1800 rpm, there is no voltage induced in the rotor winding and the motor develops no torque. As the rotor slows down, more and more current is induced in the rotor winding by the slip frequency between the line and the rotor, and the motor develops progressively higher torque. This ability to slow a little and then develop more torque makes the induction motor a very useful type and accounts for its high popularity in home appliances. The curve of Fig. 3-6 illustrates the speed/torque curve for an induction motor.

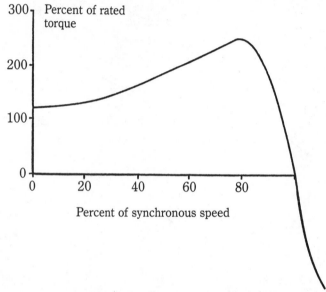

**3-6**   A curve illustrating speed versus torque for an induction motor.

The induction motor of Fig. 3-5 will continue to run very well if it has first come up to speed and the voltage is removed from one of the windings, CD or EF. Once it is near operating speed the flywheel effect will keep it synchronized with the now pulsating but *not* rotating magnetic field. At the start, if only a single winding is energized this motor develops no starting torque. All induction motors start as polyphase devices. Once started, they may continue to run as single-phase devices. For some motors, the phase shift for the other winding is obtained by placing a capacitor in series with the winding. For the more common type, the phase shift is obtained from the design of the winding. In this case, however, the design of the starting winding is such that it cannot be continuously left across the line without burning up. To protect the starting winding, a centrifugal switch is provided to interrupt the current to the starting winding when the motor exceeds about 65 percent of the synchronous speed. It is this winding that usually burns in an undervoltage condition. If the motor does not get up to the trip speed, then this switch does not open to protect the starting winding.

# Starting characteristics

Nearly all motors have the same basic pattern of starting characteristics. With the motor at rest, it is developing no counter emf to buck out the supply voltage. In addition, the induction motor might have a starting winding shunted across the line. When power is initially applied, the motor draws a big slug of current and begins to accelerate up to operating speed. As the motor picks up speed the counter emf rises and the motor draws less and less current. At about 65 percent of synchronous speed, the starting winding will drop out and the current will fall further. At this point the current drops more slowly until the motor has reached an equilibrium where the counter emf is bucking out all but enough current to supply the reactive and real power requirements of the motor.

For the 1/3-horsepower ball-bearing motor used on the drill press referenced in Fig. 1-3, the actual current figures are:

| | | |
|---|---|---|
| Initial starting current | 19.1 amperes | 78° phase |
| 1400 rpm (before starter dropout) | 12.0 amperes | 65° phase |
| 1400 rpm (after starter dropout) | 9.0 amperes | 58° phase |
| 1750 rpm (no-load speed) | 5.0 amperes | 84.5° phase |
| 1620 rpm (full rated load) (1/3 horsepower) | 5.5 amperes | 68° phase |
| 1440 rpm (stall torque 0.73 horsepower) | 6.0 amperes | 38° phase |

**Note:** In the last condition the motor is severely overloaded and will burn out in 5 to 10 minutes from overheating. If the motor passes the torque peak on the low side, the speed will fall even more sharply and the starting winding will cut in at 1200 rpm. In this case the current will again jump to 12 amperes, and the motor will burn out in a minute or so if it is already hot from overload.

It is noteworthy that for all the above figures we could have kept the line current below 5 amperes in each case with the use of appropriate power factor compensation. You will note that the compensation would have to be dynamically changing since this action is quite fast. With the light load of the drill press, which spins very freely when not actually drilling, the motor will come up to speed in about three seconds and it drops off the starting winding at about two. However, the motor on our automatic washer is another matter. It can take up to 15 or 20 seconds to drop off the starting winding when it is accelerating a full rated load of clothes and water on the spin cycle. This begins to tax the rating of a 15-ampere fuse if there is anything else on the line simultaneously.

For the small self-sustaining electrical system, the thing to look for in establishing the required short-term rating of the system is the current draw of the appliances that have a large inertial load to run: the automatic washer that must accelerate a heavy tub full of water and clothes, the furnace blower that has to accelerate a large and heavy squirrel-cage fan, the refrigerator, etc.—these are the items that tend to draw a very heavy current during the starting cycle.

The capacitor start motor and the capacitor start/capacitor run motor usually represent a somewhat better power factor than the usual starting-winding-induction motor. These units also tend to be more expensive. In certain applications such as the furnace blower it is sometimes possible to purchase and substitute one of these units. However, in the majority of household applications, such as refrigerator, automatic clothes washer, and automatic dishwasher, the motor is installed as an integral part of the unit and the substitution of another unit with a better power factor is not practical.

# The incandescent lamp

Before leaving the subject of starting surges it is worthwhile to note that most resistive heating elements change resistance considerably in the course of heating up. The 100-watt lamp on the drill press, for example, will draw an initial current of 10.8 amperes at the first instant that it is connected across the line. This rapidly falls to the operating current of 0.82 amperes rms as the lamp lights. This starting surge is very brief, lasting only a few cycles; however, it is enough to make a small generator "hiccup," and the starting inrush must be considered if an electronic switch is used. A ratio of about 12:1 for inrush surge to operating current is not unusual for lamps, and the lamp resistance actually changes during the cycle since the lamp tends partially to cool when the voltage is in the low part of the cycle. This accounts for the wave distortion seen in the oscillogram of Fig. 1-3.

Other heating elements such as irons and toasters also show an inrush surge. However, the ratio of inrush to operating current tends to be much lower—on the order of 2 to 4—the elements do not operate nearly as warmly as a modern incandescent lamp filament. The inrush current tends to last much longer since the device usually takes several seconds to come to incandescence.

# Magnetization

Items such as transformers (and to some extent induction motors), which operate by having the line current magnetize an iron core, are subject to another inrush phenomenon. If the core was last operated on, say, a positive-going cycle there will be some residual magnetism left in the core. If the next turn-on occurs on the same positive-going part of the wave, the core will usually saturate and the current will reach very high values limited mostly by the winding resistance during the first half cycle. This effect is very brief and must be considered mainly in terms of the half-cycle surge rating of electronic switches. The effect is worst in transformers that were operating without load when they were turned off.

# Rectifiers

There are a certain number of electrical devices that operate using only one-half of the cycle. For example, some inexpensive speed-control drills will use a single silicon controlled rectifier (SCR) for control of the motor. Also some heat control devices will operate by supplying only half-wave current when the device is up to temperature and full-wave current when the device is below temperature. This yields a net dc term in the line current. The net dc term tends to bias the magnetism in the alternator or an inverter transformer away from zero and can cause saturation of the device in one-half cycle. This is particularly hard on small alternators and inverters, and any condition in which a significant dc component develops should be avoided. These subjects will be discussed at greater length in the chapter dealing with power factor correction and load control.

# Generators and alternators

Under the right circumstances, nearly any machine that will operate as a motor can be made to operate as a generator, also. If the shaft of the device is simply turned faster than it would run as a motor, the counter emf will rise to exceed the line voltage and the machine will supply power to the line.

The machine of Fig. 3-5, for example, if supplied with a dc excitation on the rotor AB, will supply an alternating voltage at terminals CD and EF. If the rotor is turning 3600 rpm or 60 revolutions per second it will supply a 60-Hz voltage at these terminals. As noted before, this is a two-phase machine in this configuration.

On the other hand, if the machine were to be excited with a dc current on windings CD and EF so that the fixed poles were magnetized N-S-N-S and the machine was rotated at 1800 rpm, the rotor winding AB would be excited with a 60-Hz voltage. In this case, the machine would be a single phase machine. As noted earlier, if the rotor were fitted with a commutator, the output of the brushes would be a pulsating dc. For a practical dc generator there would be a multiple set of windings on the rotor (or armature) to keep the pulsations down. On the other hand, if the current were taken off the rotor with a pair of slip rings, this would be a practical alternating current machine.

The control of the output voltage is obtained by means of the field excitation. At a fixed speed and load, the output voltage is a direct function of the field excitation up to the point where the iron begins to saturate. If the field current is doubled, the output voltage doubles. On most practical machines, the field saturation will begin at 10 to 20 percent above the rated voltage output of the machine.

As noted earlier, the armature reaction can have a substantial effect on this field, particularly in an ac machine. A very inductive load can make the output voltage sag badly. Since this is in the direction of reduced magnetism in the field, the voltage can be brought back by increasing the field current up to the point where the power dissipated in the field winding begins to overheat the machine. As a practical matter, this means that the machine will have only a limited ability to handle badly lagging loads.

In the capacitive load case, the tendency for the voltage to climb means that the field current must be reduced. It is not unusual to find that there is enough residual magnetism in the field for the machine to climb right up to saturation with no field current at all. Theoretically, it would be possible to supply a small amount of reverse excitation to cancel partially the residual magnetism in the field and bring the voltage back to normal. However, this is a delicate operation and probably not practical. As a practical matter it is usually better to prevent the machine from ever looking into a very large capacitive load.

In a home electrical system the existence of a large load with a leading phase angle would be very unlikely except for the fact that we will be introducing leading phase components for power factor correction. If the inductive load for which these components are inserted is suddenly removed, then the capacitive elements must be quickly removed in order to prevent the system voltage from soaring.

It is not widely appreciated that an induction motor can also be used for power generation and that it has some unique properties. If you will refer back to the curve of Fig. 3-6 you will see that at shaft speeds above the synchronous speed, the induction motor output torque is shown going negative. This means that the motor is absorbing torque from the shaft and that it is actually delivering power back into the line. Unlike the machine of Fig. 3-5, which must be operated at precisely 1800 or 3600 rpm in order to produce 60-Hz current, the induction machine will produce current precisely in step with the line over a range of speeds. Typically the machine would be running around rated load at a speed of about 110 percent of synchronous or $1.1 \times 1800 = 1980$ rpm. Useful output would be delivered from about 1900 to 2000 rpm, a variation of 5 percent.

In some cases, this speed tolerance is a very useful feature since it is not always possible to control the shaft speed of a prime mover such as a windmill or a waterwheel to much better than this tolerance on a short-term basis. A second advantage is the fact that the induction motor is the cheapest and simplest machine to acquire in a given rating.

The use of the induction motor in a generating capacity has one very substantial drawback. It can never be the only source of power in the system since it requires the presence of an alternating line voltage at its terminals *before* it can begin producing any electricity. This means that it can never be more than a helper in the system. A second and more subtle drawback is the fact that some rather special instrumentation is required to determine whether the machine is actually putting power into the system or extracting power from the system. Despite these factors, the ability to operate the induction machine as a helper without maintaining precise speed and phase synchronism with the remainder of the system is an advantage that bears consideration.

It should be noted that, once brought into synchronism with the line, a machine as shown in Fig. 3-5 will develop a powerful motor torque if it begins to lag and will assume most of the load and tend to slow if it begins to lead. Within limits, this action will tend to keep two machines synchronized once the initial synchronism is obtained. A wound rotor induction motor can usually be converted for generator usage.

# 4

# Storage batteries
# and solar cells

IT IS NEARLY ALWAYS NECESSARY TO HAVE SOME FORM OF ELECTRICAL
energy storage for a home electrical system to provide energy for the various operations that might be required when the main system is not operating. As a matter
of fact there might be a requirement for some electrical energy to enable the main
system to start up at all. You might need lamps for illumination and electricity to
operate controls, pumps, and other items just to get the main system going. This
seems like a good point in the book to get into a discussion of these systems since
the storage battery-alternator-voltage regulator represents a simple self-sufficient
system that is easily understood compared to some of the more sophisticated systems that will be required to operate the entire household.

Despite all the effort that has gone into the development of alternate electrical
energy storage techniques in recent years, the automotive lead-acid storage cell
remains the most cost-effective and viable mechanism for storing electrical energy.
The economics of scale probably have a great deal to do with this. In a given year
more than 10 million automotive storage batteries will be manufactured in the US.
The nickle-cadmium cell and the silver-zinc cell and even the venerable iron-acid
Edison cell might demonstrate advantages in terms of longer cycle life, increased
energy storage density, and sealed construction. However, if any significant storage capacity is contemplated, then the lead-acid cell comes out the clear winner in
terms of reasonable first cost and perhaps life-cycle cost. If a long life cycle is contemplated, it is probably worthwhile to price the Edison cell competivitively. However, in terms of ease of replacement and availability, the lead-acid cell will
probably still show an advantage. It is noteworthy that most of the manufacturers
who provide standby power systems for computers make use of lead-acid storage.

A typical automobile battery selling for $50 will be rated at something like 60
ampere-hours and will generally be guaranteed to at least four years of automotive
use. The 60 ampere-hours implies $60 \times 12 \text{ V} = 720$ watt-hours of electrical energy.

**49**

If we presume that we would like to operate for 24 hours on battery-stored electricity at a full 1.3-kW rate we would require

$$1,300 \text{ W} \cdot 24 \text{ hrs}/720 \text{ Wh} = 43.33 \text{ batteries.}$$

Now 44 batteries would cost us $2200. Furthermore, we would expect to have to replace the batteries after 4 years. If the money was borrowed at 12% interest we would have monthly payments of $57.71. At $0.84 per kWh the 1.3 kW $\cdot$ 720 hr/mo $\cdot$ $0.84 = $78.62. In other words, the replacement cost of the batteries for only a 24-hour backup would run about 2/3 of the cost of the electricity purchased from the power company.

Of course, when purchasing batteries in lots of 50 one could probably obtain a better price but the point remains that electrical energy storage is expensive. This factor has been one of the major drawbacks to the widespread commercial use of intermittent power sources such as wind and solar power. The electricity must be sold when it is made since there is no practical mechanism for storing it. On the other hand it might not be practical to shut down or bank a thermal plant for only a few hours of service from the intermittent source.

Returning to some of the technical characteristics of the lead-acid cell we find that some of these are very well suited to the household energy storage problem. One of these is the fact that the lead-acid automotive battery is capable of sustaining tremendous overloads. When cranking an automobile engine, the battery is capable of supplying currents of 150 amperes when it is in good condition. It can also store the energy for periods of weeks or months with relatively little diminution of the stored quantity. Toward the end of the 4-year life, these statements must be revised downward. However, we see that the battery bank is more than adequate for absorbing starting surges and performing the standby function for which it is intended.

On the negative side of the ledger, there are a number of substantial considerations. First and foremost is the fact that the electricity is stored in the wrong format. It is dc rather than ac. Connected in series, ten batteries can provide 112 Vdc. This source would be suitable for operation of incandescent lamps, certain power tools that have universal (ac or dc) brush motors and the like. Unfortunately, however, it is not at all suitable for the operation of most common household appliances. Connection of nearly any induction motor device such as a refrigerator, washer, sump pump, automatic clothes washer, or TV will bring about the rapid and dramatic demise of the device in question. Certain of these items, built for operation in yachts or campers, can be purchased for operation on 12 V or 24 Vdc. However, they are generally small and very, very dear. A far more practical solution lies in the conversion of the dc to ac at 115 V and 60 Hz so that ordinary home appliances can be made to operate from the prime storage medium. A subsequent chapter deals with this conversion.

The second drawback stems from the construction of the battery. The electrolyte used in these units is sulfuric acid. If the units are not sealed, the acid and acid fumes are continually vented to the atmosphere. This results in the rapid and destructive corrosion of terminals, fittings, racks, and all manner of accoutrements surrounding the installation. In a fixed installation with a number of batteries, a fairly substantial amount of maintenance is required to wash down the batteries and neutralize the condensate with baking soda or some similar substance.

The third and perhaps most serious problem with a large battery bank is the fact that the batteries will generate a substantial amount of hydrogen gas under certain circumstances. Beneath the hood of an automobile, this gas is rapidly blown out to the atmosphere by the action of the engine fan and the wind. However, in a fixed battery bank, means must be provided for rapidly ventilating the area and preventing the pocketing of hydrogen under the ceiling.

**Caution**    A large battery bank is a definite no smoking area. Always provide plenty of ventilation. Never use brush motors or other spark producing devices in the battery bank area. Do not install a battery bank within the home.

# The generator

For many of the same reasons that the lead-acid cell is the logical choice for electrical storage, the automotive alternator represents a relatively logical choice for the prime generation of electricity to charge the batteries. To begin with, the alternator is made in such vast numbers that the economics of scale tend to reduce the price below that of nearly any comparable machine. Added to this is the fact that the auto graveyards gracing our countryside represent a vast repository of these machines, awaiting recycling. If the home electrifier is so disposed and willing to remove the unit from the wrecked car himself, these units cost from $5 to $10 with perhaps the voltage regulator thrown in. If the unit is purchased as removed by the salvage operator, the price is likely to be closer to $25. The automotive alternator is too much of a bargain to be overlooked.

The automotive alternator is a truly flexible machine capable of satisfactory operation over a wide range of speed and of remarkably robust construction. It will with equanimity survive the heat and vibration of the engine compartment and dousings with salt water. A typical rating will be 14.2 Vdc output at 40 amperes or 568 watts. At this rating, it will absorb something on the order of 0.8 horsepower from the shaft. Three such machines are capable of providing our average 1.3 kW load and five or six, with the backup of the battery bank, would easily meet the normal peak/average ratios and keep the battery bank fully charged in the bargain. The machine takes full advantage of the economies of high frequency but outputs dc so that the problems of synchronization, phasing, and high frequency losses are eliminated. The illustration of Fig. 4-1 schematically depicts this machine.

In a typical automotive installation, the 12-pole alternator rotates at twice the crankshaft speed and thus turns up something like 1200 rpm while idling and 6500 rpm at 60 mph. It is driven to speeds like 10,000 rpm at maximum engine speed. The view of the rotor assembly shows the rather unconventional shape of the pole pieces, and the armature assembly. The field winding is a straight spool of wire, which is easy and inexpensive to wind. The pole pieces themselves are fluted cup-shaped affairs that are press-fitted together and almost completely encapsulate the field winding. This design has the advantage of providing the very high centrifugal bursting strength required for such a high-speed machine while at the same time

**4-1**   The automotive alternator.

lending itself to the extremely low-cost, automated assembly required of automobile components.

The three-phase stator is wired through a three-phase bridge rectifier, which serves two purposes:

- It rectifies the ac output of the stator for battery charging.
- It prevents the battery from discharging back through the field and stator. This latter point eliminates the use of a high-current relay that used to be required.

The very wide speed range available with this machine can be a marked advantage in home electrical installations since it makes possible the use of prime sources that have very poor speed regulation. For example, a windmill with a fixed pitch will tend to rotate at a speed proportional to wind velocity. Operation of a synchronous alternator would require some mechanism for keeping the speed of the alternator at precisely synchronous speed. With this machine the speed can be allowed to vary at well over an 8:1 range.

From the very earliest Kettering self-starter installations, it was recognized that automobile engines operate over widely varying speed ranges and that the provision of an electrical mechanism for regulating the generator output voltage would be considerably cheaper and more reliable than a variable-speed transmission for the generator. The transistor/zener arrangement at the bottom of the illustration is a simplified representation of a modern solid-state voltage regulator.

The usual six-cell lead-acid battery in the car develops about 12.2 volts at rest and should not be continuously charged at voltages much higher than 14.2 volts. If the battery internal resistance is 0.1 ohm, the battery charging current will be approaching (4.2 − 12.2 volts) ÷ 0.1 ohm = 20 amperes at this level (actually the current will run a bit lower). However, since we have noted that the $V_g$ is directly proportional to $\omega\mu$, then in a speedup from 12.3 volts at idling to top speed, the alternator output will rise by a factor of 10,000 ÷ 1200 = 8.33 to a level of 102.5 volts. Accordingly, it is necessary to provide some means of weakening the field by a factor of about eight in order to preserve the battery.

A simplified voltage regulator is shown in the schematic at the bottom of the figure. Diode ZD1 is arranged to go into conduction at about 12.5 volts. With the motor stopped or running slowly, the only current through R3 is the base current of Q1. Diode ZD2 serves to offset the voltage at the base of the transistor above point B and the transistor saturates with a current controlled by R1 and the field drop plus the $V_{sat}$ of the transistor. This provides maximum field to the alternator. As the alternator turns, the voltage at point A will rise and ZD-1 will go into conduction, which raises the voltage at B, thus reducing the forward bias of Q1. The current at A is actually a humpy 3-phase rectified affair; however, the voltage is largely smoothed by the battery. Nearly the entire voltage increment will appear at point B, which will drastically lower the collector current of transistor Q1. This in turn lowers the field current in the machine enough to control the output voltage of the machine to 14.2 volts over the wide shaft speed range. The diode CR-1 is a "freewheeling diode" placed there to protect the transistor from the inductive surge that would result if Q1 were suddenly cut off by a line surge.

It should be noted that in this regulator transistor Q1 is operating in the linear mode and that it will be dissipating a power equal to the field current times the voltage drop across Q1. This means that it must have an adequate heat sink. Naturally, if you intend to use the alternator in the manner that it was used in the auto, you can simply use the voltage regulator supplied with the unit initially and the details of the operation are not too important to you. On the other hand, there can be certain advantages to building your own regulator, which we will discuss shortly.

# Hot rod operation

Since we have seen that the automotive alternator is capable of providing a 14.2 Vdc output at 1200 rpm, that the field current does not overheat the machine at this setting, and that the machine is capable of operating at a much higher shaft speed, one natural question arises. Could the machine be operated at 28.4 volts or 36.6 volts by simply turning the shaft at 2400 or 3600 rpm and making provision for a different voltage regulator? The answer to the question is probably "yes," with certain limitations.

It must be remembered that these machines were constructed to be sold in one of the most fiercely competitive and cost-conscious businesses in the world. It is unwise, therefore, to depend upon the existence of safety factors that are larger than required to make a reliable product. We know that the machine will safely run for protracted periods at speeds in excess of 7000 rpm. Therefore, the speed per se is not a problem. However, in normal usage, the voltage regulator knocks the field excitation down to provide an output of 14.2 V to protect the battery. Suppose that we decide to let the field excitation rise so that we can charge two batteries in series or perhaps three batteries in series; what happens to the machine?

The first point to examine is the full-wave bridge rectifier. On most machines this consists of a series of diodes that are pressed into the stator structure for heat sinking. These diodes might or might not have an adequate peak-inverse voltage rating (PIV) to permit such operation. Some machines will use a GE model A44F or A44A diode in each of the six bridge positions. If these can be driven out and A44B or A44C diodes substituted satisfactorily, bridge operation can usually be assured at the higher voltage.

The second question concerns the turn-to-turn insulation and the turn-to-frame insulation. In general, this seems adequate for the higher voltage operation. However, an older machine with signs of heavy corrosion might be a good one to avoid. Careful cleaning and electrical varnishing of the coils can help this situation.

The third question concerns the shaft and pulley. Since the machine will be delivering two or three times as much output power, the shaft must accept two to three times as much input power as the unit does at 14.2 V output. The torque input to an alternator is due mainly to the current drawn from the machine and to some extent to windage losses. Windage rises about as the cube of speed; however, since it starts rather small, it will probably not be excessive but the combination of higher speed and torque could reduce shaft and bearing life. Since shaft and bearing failures are not very common with these machines, though, the unit will probably operate satisfactorily from a mechanical viewpoint.

Because of the variety of alternators used on different cars, it is unwise to make too many generalizations. If the machine is bought new, the manufacturer may be queried about the operation under "hot rod" conditions. In the more likely case that the machine is purchased used, it can be tested. While I have actually operated a few machines at higher voltage without alteration, I would not recommend this for any application where reliability is of any concern. Probably the best bet would be to replace the bridge diodes and to clean and varnish the stator windings in any event. Even with this attention, there is little doubt that the higher voltage operation does not enhance the reliability of the machine.

# System connection

There is obviously something missing from the previous discussion. We have something like 25 batteries and 3 to 6 alternators but we have not discussed a mechanism for connecting these items together. An obvious mismatch exists between our generating capability and our storage capability. How do we best connect these items into an integrated system?

There are a very large number of ways in which the system connection could be accomplished. If we wished to have the facility of placing any battery in the system in any location, this could be accomplished with a crossbar switching matrix as sometimes used in telephone exchanges. However, this would require the use of some 1250 switches for the batteries alone. This is obviously not a very effective way of achieving minimum system cost. Therefore, let us examine some of the basic features that we might like to obtain from the system in order to provide maximum flexibility at minimum system cost.

First of all, the system probably should provide something more than 12 to 14.2 V output since the inverter will generally function more efficiently on a higher output. Secondly, the system should be capable of supplying some surge capability. However, this should be limited. If all the batteries were simply wired in parallel, the system would have the capability of dumping something like 3750 amperes into a short circuit. A current of this magnitude will fuse a No. 1 B&S gauge wire in about eight milliseconds! This would be a pretty violent electrical explosion. Thirdly, it would be nice to be able to remove any battery from the system without having to shut down the system. There might be times when it was important to keep the electricity running and a "weak sister" was discovered in the bank in use.

The circuit shown schematically in Fig. 4-2 provides some of these features in reasonable measure. The batteries are organized into ranks and columns. At any one time only a single column may be connected across the string of alternators. When the string is to be changed, the second string is connected first and the first string is then removed. Probably a period of one to two hours per string will suffice to keep the entire bank charged as long as the alternators themselves are running.

It might be seen that with this arrangement, any of the busses might be grounded. Therefore any given battery need only have 12 to 14 volts with respect to ground. This voltage is low enough so that any battery may be removed from the circuit without danger while the bank is in operation, provided another column is in operation. The fifth alternator is shown as a standby unit. This may be used to keep the standby battery charged or to substitute for one of the four operational units so that the operational unit may be taken out of service for maintenance and such.

Figure 4-3 shows the suggested construction for the battery storage shelter. The use of cedar or cypress lumber in the battery area is recommended if a life in excess of 10 years is to be achieved. Similarly, we recommend stainless hardware.

**4-2**  An economical connection for 5 alternators and a total complement of 25 batteries is shown.

The battery room is constructed to take advantage of the siphon action brought about by solar heat and the prevailing wind. Fresh air enters through the screen at the bottom of the door and exits in the spaces between the roof rafters. Because of the differential velocity of the wind above the ground and the slope of the roof, the pressure at the top will be lower than the pressure at the bottom.

The switches in Fig. 4-2 may be replaced with heavy-duty plug and jack arrangements. Alternatively, if automatic column switching is desired, relays may be used in these locations. Note that all switching and possible sparks take place on the switchboard side of the bulkhead.

Before attempting to remove a battery, always make sure it is disconnected from the circuit. Always leave the door open for a few minutes before entering the battery room. No smoking in the battery room.

# Solar cells

One of the most persistent questions concerning home energy projects is the question "Why not just put solar cells on the roof?" By now everyone is familiar with the pictures of satellites with their solar panels spread in the sunlight. The solar cell is silent, has no moving parts, and should last nearly forever. Direct conversion of sunlight into electricity would be great.

If one constructs a big flat diode from a semiconductor with a PN junction and one side transparent to light or infrared radiation, one has a potential solar cell. If photons with an energy greater than the band gap in the PN junction strike the junction, they generate electron-hole pairs. These pairs can separate and the junction becomes forward biased. A current will flow in an external circuit. Solar cells

**4-3** In the battery rack area, lumber should be cedar or cypress for maximum life. Hardware should be stainless. Battery cables fit through tight holes in the polyethylene and plywood bulkhead.

can be made from silicon or gallium arsenide. Theoretically, the efficiency can reach 15%; however, most cells run in the 4 to 5% range. Some cells have demonstrated efficiencies in the 11% range, using two junctions at two different wavelengths.

Gallium arsenide is considerably more expensive and harder to work with than silicon; however, it makes better solar cells. Unfortunately silicon is pretty expensive too in transistor grade monocrystalline form. The falling price of transistors and integrated circuits might lead you to believe otherwise; however, you can probably get 4000 saleable 64k memories out of a single 6″ wafer. You can get 1 solar cell!

In 1978, when I was working on the first edition of this text, there were a large number of people working on processes intended to make solar cells considerably cheaper. One of the techniques under study was the growing of "dendritic" crystals in a continuous ribbon. This would eliminate the sawing and polishing operation to make the wafer since the ribbon could be used directly. There are still a few people working on this scheme; however, the cheap, available solar cell has not appeared.

Another technique involved cells made from polycrystalline silicon, which is relatively inexpensive. A polycrystalline cell is not as efficient but if it were cheap enough it might be a success for at least terrestrial usage. Unfortunately these cells have not materialized either.

In 1978, the cost of solar cells was about $1.40 to $1.60 for the least expensive cells. In 1985 that price has fallen to something like $0.85 per square inch. This is still horribly expensive if one requires any significant amount of power. If we assume that we will get usable sunshine 1/3 of the time then our 1.3 kW becomes 3.9 kW. A typical cell in strong sunlight develops a voltage of about 0.5 V and a current of 0.15 amp. per square inch. This is a power of 0.075 watts so we would require 52,000 square inches of solar cells. This is an area of 361 square feet or perhaps 15 by 24 feet on the roof.

In Rochester, New York, during December through March you would have to shovel the snow off the array although you would probably not get enough sunshine to make it worth the effort. The cost of our array at $0.85 per square inch is $42,000, which is far too much to hope to recover even in areas where the sun shines nearly every day.

For the time being terrestrial solar installations will probably remain small expensive plants to power telephone repeaters and microwave radio links in remote areas where power is not available.

# 5

# Dc to ac conversion

IT WAS NOTED EARLIER THAT FOR ALL PRACTICAL PURPOSES IT IS NECESSARY to convert the stored dc energy in the battery bank into alternating current, if something other than a few lamps and perhaps some hand tools are to be operated during periods when the prime power source is idle because the sun is not shining or the wind is not blowing. In this chapter we will attempt to describe some of the ways in which the dc to ac conversion may be accomplished.

## Frequency and waveshape control

To begin with, let us discuss some of the requirements of your home electrical system in terms of the frequency and the waveshape. The power company controls the frequency of the power supplied to you within rather close tolerances. In general, the frequency measured on the power line does not depart by more than $\pm$ 0.2 Hz, measured over short periods such as a half cycle, and the stability averaged over a day is usually within one part in $10^6$. The reason for this is the fact that the systems are intertied. The Rochester Gas and Electric company will sometimes buy power from Ontario Hydro, Con Edison, or Niagara Mohawk through the New York State Power Pool. A pair of machines such as those shown in Fig. 3-5 may be tied together if they are operating at the same speed, and at nearly the same phase angle. If we presume that the dc excitation is applied to the rotor windings AB, and the output taken from the field windings CD and EF, two such machines may be paralleled by connecting E1 and E2 and F1 and F2, etc., provided the instantaneous voltages EF1 and EF2 and such are essentially identical. Under these conditions we would find that a torque curve such as the one shown in Fig. 5-1 would be obtained for either of the machines. The torque required to turn the shaft of either machine would depend upon the angular difference between the shafts of the two machines and the field excitation. If the shaft of one machine starts to lag behind,

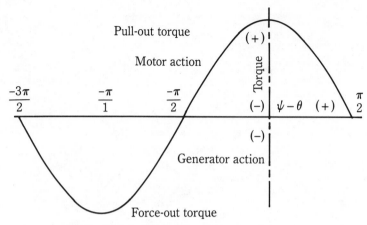

**5-1** The synchronous motor torque curve.

it will begin to act as a motor and extract power from the line. This will reduce the torque required from its prime mover, which will creep up and reduce the angle. Application of more torque to the shaft of either machine will increase the fraction of the system load borne by that machine. Within the range bounded by the pull-out torque and the force-out torque, the machines will be synchronously locked and can share the external load with the work division determined by the amount of torque applied and the excitation. If synchronism is lost, the machines will fight one another disastrously and must be immediately disconnected.

Obviously, it is much easier to provide for synchronism in a large system having many contributing machines if the frequency is agreed upon and rigorously adhered to. However, in a small system for home electricity, is it necessary to maintain such precise control? The answer is that it might be to your advantage to maintain precise frequency control for a number of reasons.

First of all, it might be of advantage to operate two machines in synchronism in a load sharing manner for a number of reasons. For example, a small hydro installation might be used for standby power to back up a wind installation. When the wind is blowing strongly, the hydro could be shut down entirely and in periods of light wind, the hydro might share the load with the wind installation. This would conserve water in the impoundment, which is usually a big consideration in hydro installations.

Secondly, since the power frequency is known to be precisely controlled, many of the appliances in the home employ the power frequency for timing purposes. The timers on your kitchen range, the microwave oven, the automatic clothes washer, and dishwasher all operate on the power frequency. The picture on your TV set is synchronized to the power frequency so that interference from the power line is minimized and the rotational speed of all but the most expensive stereo turntables is determined by the power frequency. In addition to this, nearly all the electric clocks about the home keep time by synchronizing with the power frequency. If the frequency were allowed to err by 1 Hz, the clock would gain or lose 1 minute per hour or 24 minutes per day. Therefore, it might be seen that a great many of the uses made of electricity in the home require a very precise control of frequency. In some cases such as the microwave oven, the error of one part in sixty in timing might not be critical. However, in others such as a stereo, the records and tapes will play sharp or flat by noticeable amounts. The TV may be afflicted with fluttering or creeping ghosts, which can be rather annoying to view.

Though less critical, another reason for maintaining the power frequency close to 60 Hz stems from the fact that with increasing frequency a capacitor will draw more current whereas an inductor will draw less. Circuits such as power factor correction and filtering circuits rapidly unbalance as a function of power frequency. There are a great number of such circuits hidden in modern appliances. Unless the appliance carries the legend 115 V/50-65 Hz or something similar, it is difficult to guarantee that the appliance will function on "off-frequency" power.

For these reasons, we recommend a precision frequency control to yield the most effective and satisfactory home electrical system. A more sloppy system will do some of the things you have learned to expect from electrical power, but not all. Techniques for obtaining high-precision 60-Hz power will be discussed shortly.

# Synchronizing machines

Before leaving this point, it is worthwhile to examine a simple technique for synchronizing two alternators. Figure 5-2 shows a time-honored method of determining if two machines are synchronized.

**5-2** A time-honored method of determining if two machines are synchronized.

The two three-phase machines are connected together through three large lamps, L1, L2, and L3. The lamps might have to operate on a voltage equal to twice the system voltage and should be chosen accordingly. Let us presume that machine M is operating at the system frequency and that generator N is stationary. In this case all three lamps will be lit at half brilliance. When machine N is accelerated, as it approaches synchronous speed, if all the connections are correct, all three lamps will begin to go brighter and then dimmer. The cycling will become slower and slower as machine N approaches precise synchronous speed. At this point, the field excitation on winding IJ is adjusted to give maximum contrast in the lamp brilliance. All three lamps should go completely out in the dim portion of the cycle and should reach full brightness on the bright portion. When this condition is obtained and the lamps are cycling at something slower than once per second, switch S1 might be closed the next time all three lamps are out. The synchronizing torque illustrated in Fig. 5-1 will then tend to hold the two machines in synchronism and the split of the load will be controlled by the field excitation and the shaft torque of each machine.

Sometimes on a first connection it will be found possible to make one of the lamps, say L1, go out while L2 and L3 are burning brightly. In this case the phase rotation of the two machines is reversed. This can be remedied by swapping either connections E and F or connections B and C, or by reversing the direction of shaft rotation on one of the machines. Reversal of one of the winding terminal pairs is usually easier than reversing the shaft rotation.

The use of the lamps to detect the difference in voltage between the generators on an instantaneous basis works equally well on single-phase or split-phase systems where one or two lamps would be placed in the hot leads. Never interrupt the ground lead. Fuses have been left out of the diagram for simplicity.

*Do not leave out the fuses in an actual setup.*

# Waveshape

As seen in the earlier oscillograms, the voltage supplied to you by the power company takes the form of sine waves at a frequency of 60 Hz. It is usually rated at 115 Vrms or 110 Vrms, which means that it will develop as much heat in a given resistor as 115 Vdc or 110 Vdc. The sine wave voltage actually follows the relationship:

$$E = 162.6 \text{ Vsin} (360° \cdot 60 \cdot t)$$
$$\text{for 115 Vrms}$$

where $t$ = time in seconds.

This says that the voltage actually gets as high as $\pm 162.6$ V for brief periods in the cycle and is lower the rest of the time.

Now suppose that we were simply to take a string of batteries that added up to 162.6 volts and arrange a switch to reverse the connection between the battery and the load 120 times per second. This would be a 60-Hz square wave. If this were connected to a load consisting of a 100-ohm resistor, the peak current would be equal to the peak current for the 115 Vac source. However, the power dissipated in the resistor would be:

$$\text{Power} = \frac{E^2}{R}$$
$$= \frac{(162.6)^2}{100}$$
$$= 246.5 \text{ watts}$$

For the 115 Vrms case we would have:

$$\text{Power} = \frac{(115)^2}{100}$$
$$= 132.25 \text{ watts}$$

The square wave would be dissipating 1.414 times as much power in the resistor. This tells us that the corners of the square wave contain an extra 41.4 percent of the power compared to the sine wave. Since the torque of a typical induction motor is largely determined by the peak current of the sine wave component, we can see that we would be putting a good deal extra into the motor without getting much extra out. This is somewhat of an oversimplification, but it serves to illustrate the point. In general, the corners of the square wave can be shown to consist of higher frequency components, odd multiples of the 60-Hz frequency. Because the laminations of the typical motor and transformer are too thick, frequencies above 60 Hz tend to be dissipated in eddy currents flowing in the laminations. The cores can become very hot, frequently hot enough to burn out the component.

If we were to reduce the battery voltage to 115 Vdc, we would make the power dissipated in the resistor equal to the ac case. However, the output voltage of a transformer-rectifier combination as used in a stereo or TV would be only 71 percent of normal and the torque of an induction motor would be only about 60 percent as great. Furthermore, both units would still experience excessive heating. Lamp bulbs and some universal hand tool motors would fare better.

In order to operate successfully the full spectrum of home appliances, it is necessary to provide a good sine wave output. This is a point of particular concern if one of the low-cost inverters sold at radio stores is considered. In order to operate with minimum iron in the transformer (thereby saving cost and weight) these units generally produce high-frequency square waves. These units will generally not operate anything with an induction motor and will in fact usually destroy themselves or the appliance if left connected to one for very long.

# The motor generator set

Perhaps the easiest and most straightforward way to convert dc to ac and produce good sine wave output is to use a dc motor to turn an ac alternator. If the motor speed is controlled to turn the alternator at precisely synchronous speed, the system can very nicely provide 60-Hz sine waves of good waveform. The qualifier "good," rather than "perfect," is used here because it is difficult to design small machines to provide an exact sine wave output. However, a well-designed machine can be built to have no more than 1- or 2-percent harmonic content.

Motor generator sets can be purchased in a wide variety of styles and ratings, including machines that are packaged in a single frame with no shaft output. However, for purposes of generality, the arrangement using a separate motor and generator with the shafts tied together through a flexible coupling is shown in Fig. 5-3.

In this arrangement a dc motor provides the turning shaft to power the ac alternator. In this case the alternator is assumed to be a two-pole 3600 rpm machine. Hanging across the alternator output is a device marked *output voltage sense*. This device senses the output voltage and drives a field current control for the alternator. If the output voltage of the alternator sags because of a heavy resistive load or an inductive load, the field current control increases the field current enough to bring the output voltage back to the desired 115 Vrms.

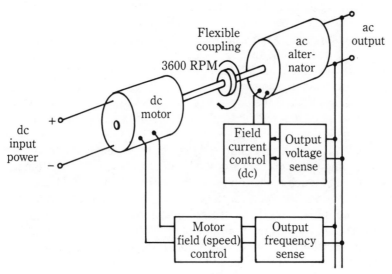

**5-3** The motor-generator set.

The second sensor across the output line senses the output frequency of the alternator. If this frequency is low, the motor speed control *reduces* the field to the motor, which makes the motor draw more current and run faster. This point bears a little explanation. The current drawn by the dc motor is a function of the difference between the line voltage and the motor counter emf and the resistance of the motor windings. When the field is reduced in the motor, the counter emf falls and the motor draws more current, which causes it to pick up speed to find a new equilibrium point. Conversely, raising the motor field increases the counter emf and causes the motor to draw less current and to slow down.

While it is not possible in a text of this size to provide a definitive treatment of items like motor and generator regulators and controls, it is perhaps instructive to look briefly at some of these items and in that way obtain a feel for the operating principles.

Figure 5-4 shows a simple voltage control for an alternator. An ac line transformer develops an isolated voltage proportional to the ac line voltage. This is rectified in a full-wave rectifier and applied to a three-terminal voltage regulator. These devices will develop a voltage that is nearly independent of load and temperature. The op amp compares a sample of the rectified signal with a 7.5 V reference derived from the regulator output. If the rectified sample exceeds the 7.5 volts, the output voltage of the op amp falls, reducing the current through the Darlington pair Q1 and Q2. This in turn reduces the field in the alternator and the alternator output voltage. The voltage control may be used to set the level of the peak ac line voltage.

The gain control serves a somewhat different function. Because this is a full closed-loop servo system, it is entirely possible for the system to swing into a low-frequency oscillation. The gain control should be operated at a gain level where this does not happen.

**5-4**  A solid-state voltage control.

Diode CR1 is employed to give the alternator magnetic field a "free-wheeling" path in which to collapse without destroying the pass transistors Q1 and Q2 in the event that they become cut off. CR1 is normally back-biased. Note that the pass transistors can be operating in the linear mode and thereby dissipating a good deal of power.

The voltage at A actually has a certain amount of ripple caused by the charging to replenish the rectifier filter. If the gain of the op amp is set high enough, this can actually force the unit into a switching mode operation and the collector voltage of Q1 will be a rectangular wave at 120 Hz. Operating in this mode, the control will be duty-factor modulated and the dissipation in the pass transistors will be much smaller. This might or might not work well depending upon the machine. To eliminate this action, the gain control may be shunted with a capacitor to reduce the gain of the op amp at 120 Hz. A capacitor at the sample tap point will perform the same function.

# The speed or frequency control

To begin with, let us consider the properties of the driving dc motor to obtain a feel for the amount of control effort that is required. We shall consider the possibility of controlling both the field of the motor and the armature with the same machine. As noted earlier, if the output load of the machine is increased, the machine will want to slow down. To prevent the slowing, which would change the operating frequency, the motor must draw more current. The current that the motor will draw is given by:

$$V = CEMF + (I \cdot R)$$

where:

$$I = \text{armature current}$$
$$R = \text{armature resistance}$$
$$V = \text{supply voltage}$$
$$CEMF = \text{counter EMF of armature.}$$

Program 5-1 has been included to permit a rapid investigation of the effects of these and other motor parameters on the system response.

### Program 5-1. Motor calculator field control, by Jack Kuecken

```
100    REM  ***********************************
105    REM        MOTOR CALCULATOR
107    REM          FIELD CONTROL
110    REM        BY JACK KUECKEN
115    REM            01/23/86
120    REM  ***********************************
130    HOME
140 PI = 3.14159265
150    INPUT "ENTER MOTOR VOLTAGE ";V
160    INPUT "ENTER MOTOR RESISTANCE IN OHMS ";R
165    INPUT "NO LOAD ARMATURE CURR. ";I
170    INPUT "NO LOAD SPEED ";S
180    INPUT "MAXIMUM LOAD IN WATTS ";W
250    REM  ***********************************
260    REM      THE CALCULATION
265    PRINT "***********************************"
270    PRINT " WINDAGE AND FRICTION LOSS IS  ";V * I;" WATTS"
275 C1 = V - (I * R)
280    PRINT "CEMF = ";C1;" VOLTS"
290 WT = W + (V * I)
300 I2 = WT / V
310    PRINT "FULL LOAD CURRENT IS ";I2;" AMPS"
320 C2 = V - (I2 * R)
330    PRINT "FULL LOAD CEMF IS ";C2;" VOLTS"
340    PRINT "WITHOUT FIELD WEAKENING MOTOR SPEED "
350    PRINT "WOULD SAG TO ";(S * C2 / C1);" RPM"
360    PRINT "TO HOLD SPEED CONSTANT FIELD"
370    PRINT "MUST BE REDUCED TO ";(C2 / C1)
380    PRINT "OF NO LOAD VALUE"
10000   STOP
```

Line 275 of the program solves this equation and line 280 prints the result. Line 290 adds the no-load loss to the full rated load and line 300 calculates the new armature current, which is printed in line 310.

With the full load current known, it is now possible to solve for the new CEMF.

Now the motor can obtain the new and lower CEMF via one of two routes:

1. It can slow down.
2. It can have its field weakened by the control.

Alternative #1 is unacceptable if we must maintain a phase lock with the power grid or with another machine. On the other hand, we can hold the field and the speed both constant by raising the line voltage enough to raise the current to the required level. In this case the CEMF stays constant.

Table 5-1 shows the results for a run on the field weakening case with a very

**Table 5-1.**

```
ENTER MOTOR VOLTAGE 28
ENTER MOTOR RESISTANCE IN OHMS .0046
NO LOAD ARMATURE CURR. 9.39
NO LOAD SPEED 3600
MAXIMUM LOAD IN WATTS 3500

*****************************************
*****
  WINDAGE AND FRICTION LOSS IS 262.92 WA
TTS
CEMF = 27.956806 VOLTS
FULL LOAD CURRENT IS 134.39 AMPS
FULL LOAD CEMF IS 27.381806 VOLTS
WITHOUT FIELD WEAKENING MOTOR SPEED
WOULD SAG TO 3525.95721 RPM
TO HOLD SPEED CONSTANT FIELD
MUST BE REDUCED TO .979432558
OF NO LOAD VALUE

BREAK IN 10000
]PR#1

ENTER MOTOR VOLTAGE 28
ENTER MOTOR RESISTANCE IN OHMS .0127
NO LOAD ARMATURE CURR. 9.39
NO LOAD SPEED 3600
MAXIMUM LOAD IN WATTS 3500

*****************************************
*****
  WINDAGE AND FRICTION LOSS IS 262.92 WA
TTS
CEMF = 27.880747 VOLTS
FULL LOAD CURRENT IS 134.39 AMPS
FULL LOAD CEMF IS 26.293247 VOLTS
WITHOUT FIELD WEAKENING MOTOR SPEED
WOULD SAG TO 3395.01984 RPM
TO HOLD SPEED CONSTANT FIELD
MUST BE REDUCED TO .943061066
OF NO LOAD VALUE

BREAK IN 10000
]PR#1
```

low resistance armature. We see that the armature current would climb dramatically to over 134 amperes with a very modest 3.1% reduction of the field. If the motor had slowed it would have brought the output frequency down to 58.77 Hz.

Now suppose that we had elected to hold the speed by raising the line voltage. At a constant speed with a constant field, the CEMF would hold constant at 27.9568 volts. The current would have to rise to the same 134.39-amp level; therefore, the drop across the 0.0046-ohm armature resistance would rise to 0.618 V, which, added to the CEMF, would require line voltage of 28.584 V. We see that we would have had to raise the line voltage by 0.584 volts. That doesn't sound too bad, but remember that we have to do it with a circuit that is now passing over 134 amperes! By comparison, the field current on the motor would almost certainly be less than 5 amperes. It is obviously much easier to obtain speed control from the motor field.

To make a comparison and to illustrate the importance of having a low armature resistance, the same problem was solved using an armature resistance of 0.0127 ohms. In this case we see that the machine responds less well. The speed sags enough to let the frequency sag to 56.58 Hz, and a field weakening of 5.7% is needed to maintain speed. The resistive drop in the armature is up to 1.7068 volts, giving a required line voltage of 29.59 volts if speed were to be maintained by raising voltage alone.

Another significant factor is that the armature heating loss has risen from 83 watts to 229 watts. This affects not only the efficiency of the system but also the heating of the armature and ultimately the machine rating since the power rating of most electrical machines is based upon the power output that can be achieved with a 40-degree C rise in the machine temperature. The machine with the 0.127 ohm armature resistance could probably not handle the full 3500 watts for more than 10 minutes at a spell.

The analysis here neglects the effects of brush drop and a few lesser items. Brush drop will generally run on the order of 0.5 volts per brush on a well-designed and well-maintained machine. If you are buying a used machine, one of the things to look for is a relatively clean-looking commutator without evidence of burning or arcing. If you can run the machine, even no-load, look for sparking and burning on the commutator. A significant amount of fire is usually a sign of shorted segments.

# Determining the rating

Often, when buying a used machine you will not know what the rating will be in your service. There are several ways of estimating this. If you are attempting to evaluate a used machine and cannot run it, a guide to the current rating is the size of the wire. In "buried" service, such as an armature or field winding or a transformer, copper wire should not be run at more than 1000 amperes per square inch. You can measure the wire diameter and calculate the rating.

If the machine is actually running, or you can run it at some known current, a second technique is possible. Copper wire increases in resistance by a factor of 0.00382 per degree C. If you measure the winding cold and then hot after it has had a chance to stabilize at a known current you can calculate the temperature rise due to that current. If the rise is less than 40 degrees C you can calculate the maximum rating by:

$$((\text{meas. rise})/40) = ((I \text{ meas.}^2)/(I \text{ max }^2)$$

With a dc machine it is possible to determine the output by locking the rotor with a lever of known length. The motor is excited with a known field and armature current and the locked-rotor torque measured with a bathroom scale or smaller scale. The maximum torque is given by:

$$\left(\frac{\text{I FIELD MAX.}}{\text{I FIELD MEAS.}}\right) \cdot \left(\frac{\text{I ARM MAX.}}{\text{I ARM MEAS.}}\right) \cdot \text{Lever } L(FT) \cdot \text{Force}(LB) = T$$

Horsepower is given by:

$$hp = 2\pi T \cdot 60$$
For 3600 rpm or 60 rev/sec

Power in watts is:

$$W = 745.7 \times hp$$

# Speed sensing

As we have seen from the examples, the regulation changes are so small in a reasonably efficient system—on the order of 2 to 6.5 percent—that the use of an open-loop speed correction system is probably not practical. The variations due to temperature and other factors such as changing brush drop require that a reasonably regulated system must have a closed-loop sensing system. We shall next deal with two classes of speed sensors.

# The one-shot tachometer

There is a large class of devices that I shall lump together in the class of one-shot tachometers. In essence all these devices operate on the principle of providing a fixed quantum of energy into an integrating circuit for each increment of angular rotation. The integrated output is then a function of rotational velocity (or other parameter). Figure 5-5 illustrates this device in its simplest form. The battery charges the capacitor up to a fixed level every time the switch closes in that direction and then discharges it through the meter in the opposite position. The average current through the meter is proportional to the number of closures per second.

The circuits of Fig. 5-6 illustrate a variation of this scheme in which a slotted disc is used as a photon interrupter. It breaks the light path between the LED and Q1. When Q1 is illuminated, it saturates and delivers a short spike to the base of Q2—through C3 and R3, which have been chosen to have a time constant much shorter than the shortest illumination period. Amplifier U1 is arranged as an integrator, or low-pass filter, to average out the charge transfer. This is not a true integrator in the mathematical sense but rather a low-pass averaging filter whose time

$q$ = CE amp. sec
$i$ = nq amp.
$n$ = No. of closures
per seconds

**5-5** The one-shot tachometer.

$E = k\omega$

$E = \dfrac{kd\omega}{dt}$

$C2 > C1$

**5-6** The optical tachometer and the magnetic tachometer.

constant is determined by the values of R1 and C1. If a second such active filter with a longer time constant is used as a reference, the angular acceleration of the unit is obtainable from the difference between the output terminals. In the configuration shown, the output of U1 would run between –(V +) at zero velocity and nearly –(V + /2) at maximum speed when C3R3 was nearly equal to the on time of Q1. Up to nearly the maximum limit, the unit is quite linear in output.

Also shown in the figure is a magnetic pickup made up of a bar magnet with a winding. The zener diode will transform the output pulse into a nearly square wave above some minimal speed by clamping the negative swing to nearly ground and clipping the positive swing. This action limits the output voltage to an essentially constant voltage wave, but the duty cycle is also constant. It is therefore necessary to drive a one-shot of some form to obtain a constant output pulse width. A circuit similar to the Q1 and Q2 arrangement will suffice.

Since the accuracy and linearity of the one-shot tachometer is directly related to the stability and precision of the one-shot, the use of one of the IC one-shot devices such as the N555 or the 74121 or 74122 is recommended. These devices have been carefully designed to provide a stable and repeatable output with variation in temperature, supply voltage, and so forth.

The Photo Interrupter might be a made-up assembly such as the General Electric H-11 series or else it could be fabricated from separate LED and photo transistor elements. It is noteworthy that the LED can be driven by TTL logic as shown, and the one-shot mechanism can be implemented by turning off the LED after the initial pulse.

Both types of units have the advantage of offering no friction load although the magnetic pickup can have a detenting torque. The photo transistor arrangement is workable down to the lowest speeds, including zero, but must be shielded from stray light. The magnetic pickup is inoperative below some minimum speed but can often be easily implemented by placing the pickup near an existing iron or steel (but not stainless) gear. An earphone magnet makes a very good pickup. However, suitable magnetic pickups are available commercially. Hall effect magnetic pickups, which are workable down to zero velocity, are also available. These units often feature direct TTL output.

These tachometer techniques are essentially analog in nature. They can be made to be stable and repeatable to better than one percent with a little care and can be improved to 0.1 percent with a great deal of care. This order of accuracy requires a very careful compensation of temperature variation and such in order to minimize drift.

If the system goal does not include the operation of clocks or very fine phase matching, then this form of speed sensor is probably satisfactory. However, for any operations in which clocks and timers are to be used, the employment of a phase-locked loop is probably imperative.

# Phase-locked loop controls

When a really precise speed control is required, the choice usually goes to a phase-locked loop (PLL). There are a variety of reasons for this choice but the principal one is the fact that there are few things in this world measurable with the accuracy that can be obtained in frequency and time measurements. A rather modest quartz watch will keep time to within about a minute per month, which represents a speed error of about one part in 43,200 or $2.3 \times 10^{-5}$. With a little bit of care, a nonovened 5-MHz quartz oscillator can be synchronized with National Bureau of Standards

WWV radio transmission to one part in $10^6$ or better. With the addition of an oven, or other temperature control, a stability of one in $10^8$ is obtainable. If you had a scale this accurate, you could take two successive weighings of the aircraft carrier Enterprise and tell whether the Captain had carried his sextant aboard! If you want to get really precise you can purchase a cesium or rubidium standard with a guaranteed accuracy over the long term of $\pm 3$ in $10^{-12}$, which is settable to one in $10^{13}$. For the long-term rate, a cesium clock set at the time of the birth of Christ could by now have accumulated an error of 0.0062 seconds (if it hadn't given up long ago). And one of the best parts of this is the fact that the accuracy of frequency standards can be verified against the National Bureau of Standards by radio, anywhere in the world, with equipment available to the most modest laboratories.

In order to operate a phase-locked loop it is necessary that we have some mechanism for detecting the relative phase of two waves. We shall begin with the sine wave phase detector.

# The sine wave phase detector

The circuit shown in Fig. 5-7 has been around for a long time. It was originally used by Major Armstrong as the detector in an FM radio. We shall not treat the design of this circuit since it can be found in nearly any radio handbook. However, it is worthwhile to have a look at how it operates.

**5-7** The sine wave phase detector.

The operation of the transformer T1 is such that it drives points L and M out of phase whereas T2 feeds L and M in-phase. In Fig. 5-8 the special case for inputs A and B exactly in-phase and equal in amplitude is shown. The induced components, which are designated by the primed symbols, add up to double the voltage at L, and to cancel it at K. Obviously, the rectified voltage at J will be strongly negative, and the rectified voltage at K will be zero. If the voltage at B were reversed in-phase, the opposite case would be obtained—with K strongly negative and L zero. If both diodes were reversed the output voltage would go strongly positive at the more favored terminal.

It will be stated without proof that the sum of two sine waves of the same frequency but with any phase and amplitude is always another sine wave of the same frequency. Therefore, this circuit works equally well when the voltage at either ter-

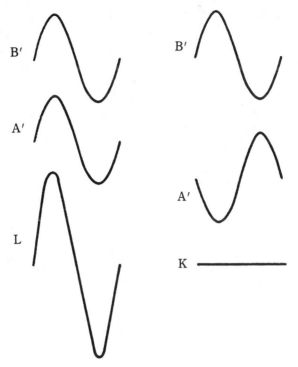

**5-8** Voltage addition at point L of Fig. 4-7 when inputs A and B are in-phase.

minal is smaller. It should be relatively easy to convince yourself that the circuit winds up with the rectified output at J and K equal when the inputs are in quadrature or have a 90° phase difference. The circuit will also work well with any waveform that is symmetrical and has equal positive and negative halves.

The latter feature is particularly handy in view of the fact that it is much easier to derive a square wave from a counter chain than it is to derive a sine wave. We can take any quartz crystal, count the output down to a suitable standard frequency, and use that for the reference applied to terminal B.

Terminal A could be hooked across the 60-Hz output of our alternator, and the voltage difference between terminals J and K will reflect the phase difference between our quartz-controlled reference and the alternator voltage.

It should be noted that the two signals to be compared should be symmetrical; that is, they must have equal positive and negative halves. This can be assured if the divisor in the counter chain always divides by an even number.

# The digital phase detector

The circuit of Fig. 5-7 requires the use of a pair of transformers, and the circuit is somewhat analog in nature. While it will work with digital signals, isn't there a way to compare the phase of digital signals? And for that matter, what is the meaning of the phase of an on and off digital signal? Let us suppose that the signals at A and B are both perfect square waves (implying a 50-percent duty cycle). Figure 5-9 shows the use of an EXCLUSIVE OR gate as a phase comparator. As noted from the truth table, this unit gives a true, or high, output only when the signals are different and not when they are alike. From this you can see that when A and B are perfectly in phase the output of the gate will always be zero, and when A and B are exactly opposite, or 180° out of phase, the output will always be one. For points in between, the unit tends to produce a pulse output whose average value is a function of the phase angle. The average value of output C is shown in the accompanying graph.

5-9   The EXCLUSIVE OR digital phase detector.

The output of a typical logic gate is typically not voltage controlled with any great precision, so the use of a voltage average of the output waveform is not a particularly accurate reference. However, EXCLUSIVE OR gates usually come more than one to a package and they *generally* will track one another pretty well with temperature (Fig. 5-9). In this case if A is inverted, it will be seen that the output of the second EXCLUSIVE OR will tend to have the opposite slope. In this case a high CMRR amplifier will be able to derive the sense signal from the pair with minimal error.

**Note:** This is a case of employing an uncontrolled parameter, namely the output one and zero voltages of a device. This is never a good engineering practice for any circuit to be produced in quantity.

For a home electrical system the violation of this rule is not too onerous and a pair of EXCLUSIVE OR gates whose outputs track reasonably well should not be too difficult to select. Because of the pulsing nature of the gate outputs an integrating op-amp arrangement such as U1 in Fig. 5-6 should be used to smooth the output.

# The phase-locked loop

Figure 5-10 shows the complete phase-locked loop in a closed-loop system. In this case you will note that a photo interrupter wheel is used instead of simply reading the line voltage. The reason for this is that the PLL must have a time constant equal to several waves before it can take any corrective action.

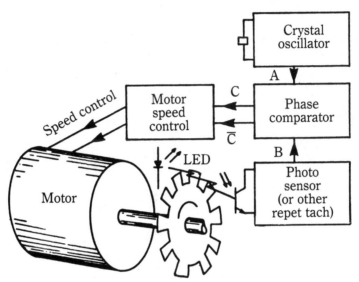

**5-10**  A phase-locked loop speed control.

As we have seen earlier, some of the starting transients (such as incandescent lamps) have a very sudden onset and are gone rather quickly. In a small electrical system, it is important that the corrective action be initiated as soon as possible. If the PLL is synchronizing to some frequency much higher than the 60-Hz output of the alternator, this rapid response is possible in a fraction of a revolution.

The photo interrupter is compared for phase with the reference signal derived from the crystal oscillator and counter chain. If the motor starts to get ahead, a signal is sent to the speed control, which calls for a slower speed, probably by reducing the voltage or increasing field current. If the motor starts to lag, the converse correction is applied.

The comparison between this form of system and the one-shot tachometer control is about the same as the comparison between a Wheatstone bridge and an ohmmeter. In the one-shot system, any variation in the capacitance or the period of the one-shot will give a corresponding shift in the motor speed. Similarly, any drift in the zero setting will be reflected as a change in motor speed. These are probably not large errors, but they can be eliminated in the phase-locked loop system. Here, the interrupter phase is continuously being compared with the phase of the crystal-derived frequency. The system has its own kinds of errors, but they can be made to be orders of magnitude smaller than the one-shot system errors.

The reduction in the time lag facilitated by operating at a multiple of the 60-Hz frequency permits the system to operate more smoothly since the corrective action will be initiated before the motor has had time to lag in speed very much. A magnetic sensor using a gear can also be used, and a veneer saw blade can make a fine-precision photo interrupter disc.

The treatment of the problem of controlling voltage and frequency has been made necessarily brief and cursory in this chapter. For a somewhat broader and more detailed look into these areas there are a number of excellent texts on phase-locked loops. There are a number of single chip PLLs on the market that combine many of the functions described here. The catalog sheets for these are also an excellent instructive reference.

# 6
# The inverter

COMPARED TO THE MOTOR GENERATOR SET, THE SOLID-STATE INVERTER would seem to hold a great deal of appeal. It has no rotating parts that wear and require lubrication. It has no brushes to wear out and require replacement and generate dirt. It has no commutator that requires periodic turning and undercutting. It furthermore should be smaller, lighter, and more convenient to use—and perhaps in the long run, it should be less expensive.

An additional line of appeal is the fact that the inverter does not require precision machine tools to build. It can be fabricated in the typical home workshop very easily. A good many radio hams have fabricated inverter supplies for their radios to permit mobile or portable operation.

With all these advantages it perhaps comes as a surprise to find the previous chapter devoted to motor generator sets and the precise control and synchronization of same. Let's take a look at the inverter and see whether we can explain some of the reasons why the motor-generator set manages to survive in the solid-state age.

To begin with, let us examine the free-running or self-controlled inverter of Fig. 6-1. This is similar to the devices one finds for sale in a radio store that will plug into the cigarette lighter in your car to power 110-volt lamps, shavers, or other light appliances. The voltage and current waveforms are shown below to assist in the explanation.

The two transistors Q1 and Q2 are initially biased ON slightly. One or the other of these will draw slightly more current than the other, and the voltage induced in the feedback winding will forward bias this unit and reverse bias the other unit: the first unit is saturated and the other is cut off. The collector current will begin to climb toward the level established by the IR drop of the winding of the ON section. As the time rate of climb of this current falls off and the current begins to knee over, the forward bias falls and the reverse bias of the other transistor diminishes

**6-1** The free-running inverter at A and voltage and current wave forms at B.

until a point is reached where the second unit begins to conduct. This rapidly switches the first unit OFF and the second unit ON. The cycle repeats for the second unit. During the OFF cycle, the cutoff transistor experiences a collector voltage equal to twice the supply voltage. The output voltage tends to be the same waveshape as $V_b$; that is, a sagging square wave.

In the case of heavier loads, the current rises faster and the switching happens more frequently; therefore, the output frequency is higher. If the load resistance is too low and the load wants to draw too much current, the feedback voltage can fall to the point where one of the transistors will no longer turn off. In this condition, with both Q1 and Q2 biased ON, the unit will simply draw a large enough current to destroy one or both of the transistors unless something is done to turn the unit off very rapidly.

This type of unit works very well when the output is rectified to dc again and it will work reasonably well as an ac supply for resistive loads. However, we see that the output voltage is a long way from a sine wave and that the frequency of the output swings widely with a varying load. Commercial units of this type will usually not operate an induction motor because the frequency is too high. Also, the high harmonic content of the wave will overheat induction motors and transformers. Large capacitive or inductive loads will usually cause the inverter to stall or overheat.

You will note that the transistors are operated in the switching mode; that is, they are either cutoff or saturated. A word on that subject is in order. When a transistor amplifier is operated in the linear class B mode, it can deliver very low distortion sine waves to the load. However, in this case the transistors can be shown to be no more than 50 percent efficient. The reason for this is that the transistors have a substantial voltage drop during the rising and falling portions of the sine wave at the same time that they are drawing a substantial current.

The disadvantages of this are not only the fact that one is wasting half of the hard-won electricity but the fact that this electricity is being wasted in the form of heat in the transistors themselves. This must be removed if the transistors are not to be destroyed by overheating. The net result of this is that the transistors can handle a far larger load in the switching mode than they can in the linear mode of operation. For example, let us consider a power tab transistor, the GE D44H11. This unit currently sells for something like $1.50 in lots of a dozen or so and is representative of moderately sized, fast-switching power transistors. The unit has the following significant specifications:

$$V_{ceo} = 80 \text{ volts} \qquad V_{ebo} = 5 \text{ volts}$$
$$\text{Power Dissipation} = 32 \text{ watts (case heat sunk to } 70°C)$$
$$V_{ces} = 1 \text{ volt ( @ } Ic = 8A \text{ and } Ib = 0.4A)$$

In a push-pull configuration, similar to Fig. 5-1 (except that the base drive would come from a sine wave source rather than the feedback winding), we see that under the stated conditions a class B amplifier could deliver a continuous output of $2 \times 32$ W = 64 watts. This is a maximum rating.

In the switching mode delivering square waves, we find that the same transistor could handle the switching of a 24-volt supply (remember, the transformer doubles the supply voltage at the collector) and when turned on, the ON transistor could pass 8 amperes. This yields a power of $8 \times 24$ = 192 watts. In this mode of operation the OFF transistor dissipates nothing and the ON transistor dissipates $1 \text{ V} \times 8$ amp = 4 watts. The division by 2 results from the fact that each transistor is ON only one-half of the time. There is also a certain amount of power dissipated in the turn-on and turn-off period since the voltage and current do not fall to zero in zero time. However, with a 1-MHz transistor at 60-Hz this extra dissipation is very small. It does not become significant until the square wave frequency gets up into the kHz region.

From the example we see that the operation of the transistors in the switching mode increases the power handling capability of a given pair of transistors by a factor of three or better and considerably reduces the heating at the same time. Unfortunately, as noted earlier, the square wave output is not very suitable for operating transformers and induction motors. This is true even if it is of the correct 60-Hz frequency because of the heating problems.

If purchasing a commercially made inverter for home electricity, examine the specifications to make sure that the unit output is regulated at 60 Hz and that the output is reasonably close to a sine wave. Harmonic content up to 5 percent is usually acceptable. Also, check the range of power factors over which operation is guaranteed. Without these precautions you might purchase a unit that was not usable for all your requirements.

From this we see one of the basic difficulties of building a solid-state inverter. The common variety of inverter does not easily supply sine waves and is very sensitive to load conditions. Both of these problems are easily handled by the motor generator set, which inherently tends to furnish reasonable sine waves and has the energy of the inertia stored in the rotors (flywheel action) to supply a portion of the momentary overloads.

It seems logical to ask whether there is not some form of inverter that can take advantage of switching-mode operation for efficiency and cool operation of the transistors while at the same time providing a more reasonable approximation to regulated 60-Hz sine waves. We shall consider a few of the possibilities.

# The quasi-sine wave inverter

One of the first possibilities to suggest itself is that the sine wave might be approximated by a suitable staircase voltage. Obviously, the sine wave can be approximated to any required degree of precision by a number of steps if the number of steps is allowed to be sufficiently great. However, from an economic viewpoint the question becomes one of deciding the number of steps that requires the smallest number of transistor switches. And yet, these steps must still provide an approximation of the sine wave that is low enough in harmonic content to avoid some of the heating effects previously discussed.

Rather surprisingly, the first such happy approximation comes with a single-step, judiciously timed. To the best of the writer's knowledge, an algorithm does not exist to permit the mathematical solution of an optimum waveform with an arbitrary number of steps. However, the harmonic content of any particular waveform is solvable by means of the Fourier Transform, which can be implemented on a computer. The waveform on Fig. 6-2 was selected from a computer search. If the waveform shown dotted is compared to the largest sine wave component it can be seen that the $V_s$ is just 14 percent higher than the largest sine wave component. If the wave period is considered to be divided into 12 equal parts, this waveshape can be obtained by turning Q1 ON during periods 2, 3, 4, and 5 and turning Q2 ON during periods 8, 9, 10, and 11.

A circuit for providing this waveshape is shown in Fig. 6-3. A 7492 divide-by-twelve counter is used to drive a 74154, 4 to 16 line decoder. This unit is an active

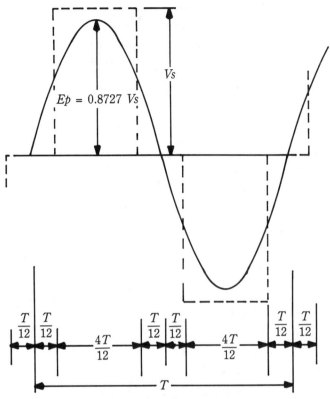

**6-2**   The one step quasi-sine wave.

**6-3**   The single step quasi-sine wave drive.

low device. Therefore, the 4 wide NAND gates are actually performing an OR function and selecting the periods noted above.

This voltage waveform has no even order harmonics and no harmonics that are multiples of three. The total harmonic distortion is 7.45 percent. This is somewhat higher than strictly desirable. However, the first harmonics with any significant content are the fifth and seventh, and some of this energy can be filtered out fairly easily.

This waveform is not quite as good as the best two-step wave investigated. For the latter the voltage is held at 0.5774 $V_s$ during periods 1, 2, and 3, at $V_s$ during periods 4, 5, 6, 7, and 8, and at 0.5774 $V_s$ for the remaining periods 9, 10, 11, and 12 of a 24-period cycle. The two-step wave has a fundamental (60 Hz) component of 0.8830 $V_s$ and a total harmonic content of 4.74 percent.

Unfortunately, the odd voltage step would require switching to different taps on the transformer and would thus need nearly twice as many transistors. If the step is selected as 0.5 $V_s$, the best waveform tested was no better than the best single-step.

Because of the significantly greater complexity and the probable reduced reliability and the question of whether all the reduced harmonic performance would be obtained in practice, it would seem that the single-step inverter might be the more practical choice. As will be seen shortly, even the simplest inverter built in a practical household size is rather impressive.

# The 2.6 kVA inverter

The design and construction of a large inverter is definitely *not* a good project to be undertaken by a rank amateur with little in the way of test equipment. If you are not reasonably well equipped and knowledgeable in electronics, the best course is to shop around and *buy* an inverter. The following design is presented mainly to give a feel for what to expect and to serve as a guide for the more experienced.

The circuit diagram of Fig. 6-4 shows the inverter circuit. In order to pass the required current it is necessary to parallel 13 of the D44H11 transistors. In the collector of each is a low-resistance resistor to prevent current hogging and to dampen spiking a bit. It is important that this resistor be noninductive. Base drive for the parallel groups comes from a PNP transistor. The base drives are equalized by the 28-ohm resistors. The PNP is in turn driven by the Darlington pair, which brings the input sensitivity down to TTL level.

Transformer T1 is an Arnold AA533 Silectron "C" core weighing 18 pounds. The secondary consists of 131 turns of B&S No. 2 gauge copper. The two primaries consist of 39 turns of B&S No. 0 gauge copper with a tap at the 33T point for operation at low voltage. The core is made of 12-mil laminations in deference to the harmonic content. If the transformer is immersed in a good transformer oil, the inverter is capable of continuous duty at 2.6 kVA. **Note:** When packing a transformer in oil it is a good idea to pull a vacuum of about 27.5 or more inches mercury to assure that all bubbles are withdrawn from the windings.

The weight of the secondary wire is about 20 pounds, and the weight of the primary winding is about 22 pounds, bringing the total weight of the unit up to

**6-4** The 2.6 kVA inverter.

about 60 pounds, including the core but exclusive of oil and container. The use of oil is probably a good idea from a cooling standpoint since it is a bit of a struggle to get the appropriately sized wire for the primary on an inverter. The primary windings should be wound bifilar and should be outside of the secondary in this design for better cooling. **Note:** *An inverter of this size produces lethal voltages. The unit should be handled with the respect given high-voltage equipment.*

The unit as shown should handle about 26 amps on a continuous basis on the output and surge to 33 amps on a one-minute basis. Because there is a continuous positive drive excitation from the driver circuit, there is no danger of stalling the inverter.

The filter on the power input is required because the unit draws its current in 5.56-millisecond slugs with a very steep rise time. With any significant impedance in the leads from the battery or generator, a great deal of ripple would appear on the battery leads. The condensers should be installed with the shortest possible leads. The choke should have sufficient air gap to permit nonsaturating operation up to 150 amps.

Care should be taken in heat sinking the transistors. Each of the large units will be dissipating about four watts maximum. The unit should be safely operable up to a heat sink temperature of 70+C.

The A-B switch in the primary is there to permit operation either from the batteries alone or from the batteries with the generator; i.e., 12 volts or 14.2 volts. This switch should only be operated with the unit nonoperational.

The pair of diodes CR1 and CR2 are present for the freewheeling function. They are necessary since the instantaneous interruption of the primary current could give very large spikes, which would destroy the transistor bank if the unit was operating without secondary load. These units should be rated in excess of 150 amperes and should have the highest switching speed possible.

# Radio frequency interference

The entire unit should be completely enclosed in a metal case with all leads coming through RFI suppressors. The construction of the case should be either all-welded (except for the entry door) or riveted or bolted with a fastener every lineal inch along the seams. Without this precaution, neither you nor your neighbors will be able to listen to radio any time the unit is in operation, and the TV and FM will also be troubled by the RFI.

# Larger transistors

Several firms currently offer transistors with higher ratings than those selected. For example, General Semiconductor Industries is offering a T0-3 packaged unit rated at 50 amps and 200 V. Starting with a 56.8 V input, these units could be used singly in the opposite sides of the inverter. Unfortunately, they are considerably more expensive, and inverters have a way of chewing up transistors during the development and test phases.

As noted earlier, the design and development of a large inverter is not a task for beginners. The internal layout of the unit should permit the use of the shortest possible leads, and the leads from any of the parallel units should be the same. Depending upon the mechanical design, the total weight of the unit with case and transformer and filters will run around 250 pounds.

A particularly tender point in a design of this type is the loss of the output load at the instant when one of the current pulses ceases. If the load has been heavy and the current in the inductor primary is built up to a high level, a large transient can result on both the input and output. Item VR1 is a GE MOV varistor intended to limit this transient. These units function very much as if they were a pair of back-to-back zener diodes. When the threshold is exceeded, they draw current heavily and tend to absorb the transient energy. For a unit of this size, the spike could contain an energy on the order of 180 joules. A recommended unit would be the GE type V13-0PAECB 20A with six parallel units required.

# The transformerless inverter

One of the significant changes in the interval between 1978 and 1985 was the emergence of power grade transistors capable of handling at least some of the requirements of a home electrical system. Operating in the single step mode the circuit of Fig. 6-5 functions as a double-pole double-throw switch to energize the load with the polarity alternating. This is a bridge function. For example, when Q1 and Q4 are on the left-hand side, the load is positive. Conversely, when Q3 and Q2 are on the right-hand side, the load is positive. The two TTL drive signals are obtained from the drive circuit of Fig. 5-3. The main switching transistors, Motorola MJ10048, are rated at 100 amperes and 250 volts so they should be easily able to handle a household load. They are actually Darlington amplifiers and should drive easily; however, the photo isolators might require a little help in driving them.

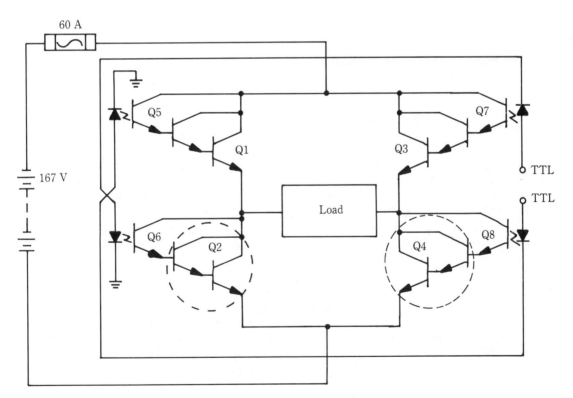

Q1, Q2, Q3, Q4 = Motorola MJ 10048 or EQ
Q5, Q6, Q7, Q8 = Phototransistor

**6-5**   The transformerless inverter.

Harking back to the battery situation, we noted in chapter 3 that it might be desirable to be able to move the ground in the battery string to any point so that we can bring an individual battery to ground potential while the system is in operation so that we might be able, for instance, to add water or service the battery without completely shutting down the system. In a transformer inverter system or a motor generator system this would not necessarily have an effect upon the load. However, with this transformerless system the ground would be established in the household electrical system by the ground on the battery. This system will not function with any ground fault interrupting breakers in the system. Your local electrical code might require ground fault interrupting breakers on outdoor and bathroom circuits. Of course you may ground the system at the load; however, this means that the entire battery stack is swinging at high voltage with each end of the string grounded on alternate half cycles. The battery cases would have to be insulated since the insulation is not intended for 167 V.

Another major problem is the fact that the ground at the load will cause some difficulties in RFI suppression. The switch would have to be kept as close as possible to the battery stack to minimize radiation and the leads should probably be shielded. The system will have to be carefully bypassed and shielded if anyone is to listen to radio or TV within a quarter-mile of the inverter.

# True sine wave inverters

A number of vendors have developed some very sophisticated true sine wave inverters for "uninterruptable" power supplies to be used with computers. The requirement for these units stems from the fact that many computers will lose the complete contents of the read-write memory with even a brief interruption of the line power. These special power supplies are built to fall back upon battery storage in the event of failure of the line voltage. Many will do this without even so much as a phase glitch on the line into the computer.

Most of these units operate at a frequency that is a relatively large multiple of the line frequency, such as 20 or 30 times. The sine wave variation is generally obtained by sine wave modulation of the duty factor either by pulse width modulation or by pulse repetition rate modulation. The high frequency is rectified and filtered to obtain the low distortion of 60-Hz sine wave.

These units tend to be very expensive, and a discussion of the design is well beyond the scope of this text. In general these units are well adapted to handling fixed loads of very high power factor. The performance of such a unit in handling heavily inductive transient loads, such as an induction motor starting cycle, is questionable. There is probably no special reason why the unit could not be built to handle such loads, but the prime application does not require such performance and the capability has probably not been designed into the unit.

# Summary

It seems to me that the motor-generator set is still attractive in applications where the load can vary widely and can present poor power factors and transients. The motor generator is inherently equipped to produce reasonable sine waves and is not inclined to be "tender" about momentary overloads. As the ratings of transistors increase and prices fall, inverters approaching the robustness and output quality of motor-generator sets will perhaps appear. At this writing, though, the edge seems to belong to the motor-generator set.

# 7

# System control and power factor correction

FROM THE DISCUSSION IN THE PRECEDING CHAPTERS IT SHOULD BE APPARENT that some very substantial economies are available without sacrificing utility if the home electrical system is rigidly controlled to prevent the occurrence of very high peak-to-average ratios and the existence of bad power factors. In this chapter, we shall discuss some of the mechanisms available to permit control of the system in such manner that these effects are limited.

Obviously, a number of simple common-sense moves can be made without any requirement for additional equipment. For example, the clothes may be dried on the line out-of-doors without *any* electric power, rather than with the electric clothes dryer. Of course, a wind, rain, snow, or sleet storm can make this impractical, and the indoor clothesline might have to be pressed into use. Another matter is as simple as making sure that all unused lamps are turned off. Dishwashers and clothes washers should be run only when there's a full load. Use the small toaster oven when the capacity of the kitchen range is not required. These simple conservation measures can save a significant fraction of the electrical energy consumption.

But there is a large class of services around the home which is not easily done without. For example, when the sump is filled with water you would like to have it pumped out. When the thermostat falls below the pre-set temperature you want the furnace to start. When the refrigerator rises above the pre-set temperature you want the compressor to start, and so on. And if you have your own well, you of course want the water pump to work.

You will note that each of the preceding examples is a machine with some level of automatic control. *You* do not normally turn it on and off; it makes that decision for itself on the basis of some internal criterion. In the home electrical system, there is little to prevent several or all of these machines from reaching the decision to start simultaneously, thus creating a very large peak demand.

This problem exists in large industrial situations also, and the utility companies actually bill not only for the average power consumption but also for the peak kW demand. In some cases, the peak kW charge can be as large as one-half of the monthly bill. For this reason a number of industrial facilities have instituted a microprocessor control system to reduce the peak demand.

Traditionally, the utilities have not billed home owners for peak kVA because of the expense of monitoring instruments and the accounting procedures involved. Instead they have simply adjusted the rates to home owners to account for the average cost. As time continues, this can change. However, at present, there is no economic incentive for the home owner to install expensive peak-load control and power-factor control apparatus as long as he is buying electricity from the utility.

In the situation where the home owner is making his own electricity, the emphasis changes dramatically. As was noted earlier with the drill press example, a proper power factor correction would have held the peak demand below 5 amperes instead of the nearly 20 required. Also noteworthy is the fact that we might intervene somewhat in the cycling of the automatic machinery so that no more than one big machine can start at any one time.

By these techniques, it seems likely that the home owner can reduce his peak kVA demand from something on the order of 15 kVA to something more like 3.5 to 4 kVA. Because the difference in price of storage-generation facilities tends to proceed somewhat faster than linearly, this would represent a reduction in initial cost by a factor not lower than 4 and perhaps as high as 12 or 16. Thus from an economic viewpoint as well as a conservation viewpoint, it makes sense to consider a fairly large investment in control and power factor correcting equipment. Because of the flexibility, the ease of usage, and the rapidly falling costs, we shall consider that one of the various single-board microprocessors will be used as the heart of the unit.

The detailed discussion of the microprocessor and its operation are well beyond the scope of this book. We shall present the various control algorithms in flow chart form only. It will be presumed that the interface between the processor and the system will be through some device such as the Motorola 6820 PIA (Peripheral Interface Adapter) chip. The detailed workings of the processor boards vary so much from one unit to a different kind of unit that we could easily fill the whole text with nothing but software for different processor chips. It will therefore be assumed that the reader knows how or will learn to program the unit he has.

# Request/permission control

Obviously, if we wish to ensure that no two large units will start simultaneously, a request/permission system is the easiest to implement since this system can be entirely external to the appliance. Suppose, for example, that each of the appliances was equipped with a voice and ears. If the temperature of the refrigerator rose above the setpoint, it could ask the controller, "May refrigerator have permission to start?" The controller could then survey the current power consumption and any other requests and reply "permission granted" or "hold." The latter response would indicate that some other machine was in the starting process. The "hold" would be in effect only during the several seconds that the other machine actually needed to start.

In this connection, remember that a number of machines have some continuing requirement for a small amount of electricity in addition to the main load. For example, the electrical timer and the panel lamps on an automatic washer draw a few watts even when the machine is in the "soak" part of the cycle.

A second thing to remember is that different machines will have different levels of urgency attached to a request to start. For example, it doesn't matter to the refrigerator or the air conditioner if it has to wait five seconds for another motor to start. Conversely, the furnace bonnet temperature control requesting a start from the blower might easily overheat in a short additional period and abort the heat cycle. Thus, priorities might have to be assigned to the system. For example, the controller might have to know that the furnace heat cycle has started and "hold" all other requests until the blower is safely started.

With these restrictions in mind, let us examine a very simple form of remote Request/Permission control. The circuit of Fig. 7-1(A) shows a schematic version of the circuitry involved. The 2200-ohm resistor in series with the line permits enough current to flow to permit operation of a small timer and such with nearly full-line voltage. During this period, the rectifier following the 6 V transformer rectifies a small signal of about 5 volts, which is sensed by the computer. When the machine attempts to start, the impedance of the appliance falls sharply and the rectified voltage on the ($\overline{RTS}$) line falls to nearly zero. This is sensed by the computer as a TTL low. If the conditions permit a start at this time, the computer takes the (PTS) line high, which closes the relay, shorting the 2200-ohm resistor, and the machine will start normally.

The selection of the 2200-ohm resistor is nominal. It serves only to limit the current and to prevent a start without the permissive (PTS) signal by limiting the current through the appliance and the ($\overline{RTS}$) sensing transformer to 0.052 amps. Since the transformer is lightly loaded, it will draw only a few milliamperes and the remainder is available to operate the clock or timer or other instrument. Note that in the case of an automatic washer, the voltage would drop far enough to stop the timer, thus "freezing" the washer in the portion of the cycle at the beginning of the "start" command from the internal logic.

In the case of automatic washers for clothes and dishes, solenoids are usually used to operate water valves. These could not draw sufficient current through the 2200-ohm resistor to operate. Depending upon the design of the particular machine, it might be necessary to bypass the (PTS) relay to permit these to operate. This will require some internal modification of the machine. Your friendly neighborhood repairman might or might not understand how to do this and whether it is required.

# Microprocessor-controlled machines

A special note of caution is in order with regard to the increasingly large number of machines that use simple microprocessors for control rather than gear-cam-motor

**7-1** REQUEST/PERMISSION TO START circuits.

timers. In general, the microprocessor requires a continuous flow of current in order to retain its memory. If the electricity is shut off—even very briefly—the processor memory will be wiped clean and the unit will go back to square 1 in the process. For such machines, it is imperative that an internal alteration be made to supply current continuously to the control logic to permit the machine to complete the cycle. Without this provision the machine will get only as far as the first high-current demand cycle and reset to the start condition.

# Solenoid valves

Another note of caution is in order regarding solenoid valves. On a number of washing-type machines, a solenoid valve is timed to turn the water ON. The turnoff command is obtained by sensing the point, after the starting surge, when the motor current either shifts phase or increases because of the load of churning the water. If the motor is not allowed to start by the load control, the OFF WATER command will not be issued and the machine will overfill or overflow. Machines using this logic generally come from older designs with mechanical clock timers. Interrupting *all* of the current will usually prevent the solenoid from turning ON, thus preventing the flood. *On all timed-cycle machines it is advisable to apply the* $(\overline{RTS})/$ *(PTS) type control only after a careful consideration of the effects of the interruption of supply current upon the cycle.*

# Solid-state switching and sensing

The circuit of Fig. 7-1(A) is shown using a relay and a transformer for sensing. It is possible to accomplish both these functions entirely with solid-state components. The circuit of Fig. 7-1(B) illustrates a solid-state ac switch with a photon isolation feature. The component designated U1A and U1B represents a packaged photo-SCR isolator such as the GE H211C. This unit is capable of operating directly from 115 V ac lines and provides a high voltage isolation to the TTL control line. The unit comes in a 14-pin dip package and can handle 0.300 amps. This is too small for most appliances large enough to warrant $(\overline{RTS})$/(PTS) control but is sufficient to permit control of SCRs or TRIACs large enough to do so. The latter should be selected to handle the appliance surge. For example, the drill press motor that draws 19 amperes on start could be handled by a pair of GE C36B SCRs.

The solid-state sensor shown in Fig. 7-1C employs a transistor photo isolator such as a GE H11A-1. The diode CR2 is any modest size diode, such as an 1N4000. It serves only to protect the light-emitting diode in the photo-isolator. These units have a very low peak reverse voltage, typically less than 5 V. It should be noted that the photo transistor isolator might not always give a good TTL level ON and it might be necessary to follow the stage with a transistor.

# Negative true logic

It will be noted that the $(\overline{RTS})$ signal is a negative-true logic signal and that (PTS) is a positive true logic signal. This is actually the advantageous way to operate the system since any of the (PTS)/$(\overline{RTS})$ devices can be unplugged from the central controller without effect. In the absence of a low on the line the controller will make no distinction between an unplugged line and a device not wanting to start and will therefore take no action.

In a similar manner, the active HIGH on (PTS) protects the unit controlled. If the unit is unplugged it makes no distinction between the unplugged condition and a HOLD, and will not start. This is important since it is possible to disconnect any unit without having anything start up unexpectedly.

# Load shedding

Most of the industrial-style load controllers function to limit the peak demand for electricity to minimize the demand charge from the utility. They generally tend to do this by classifying the various loads into SHEDDABLE and NON-SHED-DABLE loads. They sometimes assign a priority to these loads. Prime examples of sheddable loads are:

- electric hot water heaters
- sidewalk defrosters
- pool heaters
- air conditioners

Such devices can be turned off without any serious consequences except as noted in the following paragraphs.

There is usually a series of non-sheddable loads that should be permitted to operate if at all possible. This includes items such as:

- electric lights
- computers
- cash registers
- process controls

Obviously, it would not be desirable to have the lights go out or to have the computer crash every time the general load got high.

These categories really apply largely to industrial and commercial applications. For the home electric system, the likelihood that one would use his hard-won electricity for items as frivolous as sidewalk defrosters and pool heaters is very small. However, there are always some sheddable loads and also some non-sheddable loads. In addition to these, some deferrable loads and non-deferrable loads exist. Also, some of the loads may be deferrable *and* "non-interruptible."

# Load deferment

There is a class of loads that can be readily deferred but may not be interrupted, as for example, the typical forced-air gas furnace. The machine draws a small but steady load from the line to operate the thermostat system. When the room thermostat senses that the room temperature is below the setpoint, the solenoid valve opens the burner lights. The burner is usually equipped with a nonelectrical interlock that prevents the operation of the solenoid valve if the pilot is not lit. Alternatively, on some units electrical ignition is used, and the spark is ignited to light the flame when the solenoid supplies gas. On some oil units, a pump and blower are also started to furnish oil and to blow the flame.

After the flame has been burning for some time, the bonnet or heat-exchanger thermostat gives the motor start command to the blower. If the motor does not start for some reason an overtemperature interlock operates to shut off the flame. This overtemperature interlock is intended as a safety feature and is not intended to be a regular operating portion of the furnace cycle. After the blower has been operating for some time, the room temperature thermostat will pass its hysteresis range and open. At this point the heating cycle is ended.

It probably does not make a great deal of difference if the furnace electricity is shut off while the room thermostat is open. The furnace simply does not sense the closing, and the cycle is not initiated. If the start is deferred, the rooms might get a trifle cooler but nothing is harmed.

On the other hand, if the cycle is interrupted in the middle, the onus for turning off the fuel to avoid melting the bonnet is left to the overtemperature interlock. This is generally a safety control and is not really intended to be cycled at frequent intervals. It would be poor engineering practice to create a situation where the safety interlock was cycled at regular intervals. Therefore, a well-designed control algorithm would take account of the fact that the furnace load should not be shed while the furnace is in operation.

A somewhat similar situation applies to refrigeration devices. Most of these units respond to an overtemperature command by starting the compressor. The compressor usually has a pressure reservoir so that the refrigerant pressure builds slowly and the starting winding has only the inertia of motor and compressor to overcome. Once the unit is running, the pressure builds in the reservoir and the actual refrigeration process begins. When the temperature in the box passes the low setpoint, the thermostat opens and the motor stops. Because of the reservoir, the refrigeration process continues until the pressure in the reservoir has fallen to a low level.

Most of these machines do not have enough torque to start the compressor if there is pressure in the reservoir. If the current is interrupted while the compressor is running and a restart is tried soon after, the motor will stall and develop no counter emf. It will therefore draw excessive current, and the starting and running windings will overheat. A temperature sensing cutout will usually interrupt the current to prevent the burnout of the motor. When things cool down the cutout will close again and the systems will try another start. Interruption of the refrigerator might not actually destroy anything; however, it is poor engineering practice. The cutout was not designed for regular cycling, and the overheating will significantly shorten the life of the motor. Most refrigeration machines should not be interrupted while running, and if it is necessary to do so, they should be locked out and prevented from attempting a restart for a sufficient period to permit the reservoir pressure to bleed down to neutral.

# Self-defrosting machines

In passing it should be noted that self-defrosting refrigerators and freezers are notorious wasters of electricity. These machines usually apply an electric heater to cause the condensate to remain liquid and run off, whereupon they apply an evaporation fan. This obviously wastes a great deal of electricity. In a few of the machines the self-defrost feature can be disabled. For most home electrifiers, the logical choice will be to defrost the unit manually from time to time and thus save the cost and effort involved in providing the system capability and storage required for this rather wasteful practice.

The burnout type of self-cleaning oven represents an even more atrocious use of home electricity. Just do not turn on this feature unless you really intend to test the maximum load capability of your system.

# Load current sensing

While it might not be immediately apparent, an important part of the kind of load management described here is the ability to sense when the deferrable but non-interruptible loads are actually drawing current. The basic principle of the $(\overline{\text{RTS}})$/ (PTS) scheme requires the cancellation of (PTS) as soon as the device is satisfied. While this can be done in some instances on an elapsed time after (PTS), in a great many it is either not convenient or not possible to do so.

For example, in the case of the furnace, the length of time that the unit will run can be quite long in cold, windy weather. For the refrigeration devices, the time can be long in hot weather or after the door on a freezer has been held open while stocking the shelves with warm food. The overall system speed could be severely penalized by operating on a worst-case time delay. Conversely, the system can operate with maximum dispatch if the computer can sense the cessation of normal load current. This cessation will tell the computer that the furnace is up to temperature or the refrigerator is down to temperature or the sump is empty. Once the device has shut off, the computer can safely cancel (PTS) without interrupting the load cycle.

The illustrations of Fig. 7-2 show two forms of current sensors. These circuits are installed in series with the load and are represented by the ammeter symbol A in the three illustrations of Fig. 7-1. The first of these is an opto-isolator circuit similar to the $(\overline{\text{RTS}})$ sensor of Fig. 7-1(C). When the load is drawing rated current, a three-volt drop is noted across $R_s$, and the light-emitting diode of the photo isolator will be illuminated and the photo transistor switched ON. Diode CR1 protects the LED from reverse polarity. The small filter is added to ensure that the pulsations from the half-line cycle are not seen by the computer.

The rather large drop and consequent heating in $R_s$ are necessitated by the fact that it requires about 1.75 Vdc to turn the photo diode ON. This feature makes this form load current sensing less attractive than the circuit shown in Fig. B.

The circuit shown in Fig. 7-1(B) employs a current transformer similar to the one described in chapter 1. Since the voltage developed across $R_L$ is proportional to the line current, a low current will not develop sufficient voltage in the filter to bias the transistor ON.

Both of these circuits will withstand a five-times-normal starting current for a number of seconds. Both also provide a negative logic starting and running signal since unplugging of the unit will be interpreted correctly by the computer as a no-current condition, whereby the computer will cancel (PTS).

$$R_S = \frac{3\ V}{\text{Rated current\quad(A.)}} \qquad \text{(Non inductive)}$$

$$\text{Power} = 6 \times \text{rated current (A.)}$$

**7-2** Solid-state current sensor at A and transformer type current sensor at B.

# The control code

In order for the computer to be able to distinguish the various forms of treatment to be accorded each individual load, a control code or control word is established. This control code should contain the significant information. The code could be almost any length. However, an eight-bit code is convenient for most of the microprocessors used in a control application of this type. In the example that follows, this format is used.

First of all, let us assign the three lowest order bits, b0, b1, and b2, to the starting time code. These three bits can give us any binary value from zero to seven. If a unit time of 4 seconds is assigned, we can have any starting delay from 0 to 28 seconds. This is adequate for most appliances under most loads.

The next three bits, b3, b4, and b5, can be assigned according to the current draw of the device. For example:

| b5 | b4 | b3 | Interpretation |
|----|----|----|----------------|
| 0 | 0 | 0 | Less than 5 amperes |
| 0 | 0 | 1 | 5 amperes |
| 0 | 1 | 0 | 10 amperes |
| 0 | 1 | 1 | 15 amperes |
| b5 | b4 | b3 | Interpretation |
| 1 | 0 | 0 | 20 amperes |
| 1 | 0 | 1 | 25 amperes |
| 1 | 1 | 0 | 30 amperes |
| 1 | 1 | 1 | 35 or more amperes |

Next, we may assign the two most significant bits, b6 and b7, to the device classification. This is as follows:

| b7 | b6 | Interpretation |
|----|----|----------------|
| 0 | 0 | Never interrupt (reserved for lights, and other critical services). |
| 0 | 1 | Defer but do not interrupt (furnaces, refrigerators, etc.) |
| 1 | 0 | Defer or interrupt but delay restart (air conditioners, compressors, etc.). |
| 1 | 1 | Interrupt first (hot water heaters, ovens, or other always-sheddable loads). |

The purpose of this coding is to provide a scheme with a mathematically determinable value in proportion to the desirability of shedding the load. For example, a device with the code 11111 in the most significant bits (equivalent to the decimal value 31) would be a desirable one to drop in a heavy overload since it is classed as always sheddable and it draws in excess of 30 amperes.

A further value is the fact that any device classed as always sheddable will have a larger numerical value than the largest device in the defer or interrupt but delay restart category. It can be seen that the code thus organized provides the computer with a clear-cut mathematical route to select which devices are to be dropped.

At the other end of the line, at the time when the system is in the process of starting up after a shutdown, it is likely that nearly all the automatic devices will be requesting a start. In this case, the system will pick up the load in an orderly fashion by starting the devices in ascending order, building the system up to capacity, and adding items after some of the higher priority items shut off.

# The appliance number

A further item is required for control and this is the appliance number (APN). In actual fact the appliance number is simply the address in memory where the control code is written. Thus we would find the control code for device zero at x000 and the code for device number 3 at x003. This scheme is particularly handy when using a processor with an index register with offset addressing because the addresses for the code, the input and output terminals, and other requirements can be obtained from the index register with appropriate offsets.

The devices should be arranged in the order of ascending code numbers, but this scheme offers the added flexibility that devices with identical code numbers can be given a further priority by placing the one to be shed first at the higher (APN). In general, the number of devices to be controlled will be relatively small and one or two sixteen-register hexadecimal pages will suffice for the code table, with a similar number for the input/output address table.

# The control algorithm

The flow chart in Fig. 7-3 presents a control algorithm arranged to provide for orderly assumption of load and orderly restoration after a load shed has occurred. It is assumed that none of the devices coded 00xxxxxx are equipped with either switches or current sensors and that these devices will be shed only manually, as by turning off the lights and so on. The various portions of the program have been labeled by enclosing the title in an ellipse.

Any complex algorithm requires some study in order to obtain a reasonable understanding of the operation, and this is no exception. However, to try to make this one a bit easier, let us "walk through" this program on a "guided tour" and comment on the sights along the way. We will begin by assuming a "cold" start when the system has been down and nothing is running.

PROGRAM START is the reset address of the processor. At this address in the ROM (Read-Only-Memory) there are a series of instructions that initialize the processor by clearing all (PTS), flags, and DELAYS. After this, the machine loads the control codes in read/write memory. The loading might be either manual from a keyboard or from tape depending upon the facilities available. The system then sets (APN) to the first 01 coded address. This is the first device to be controlled and enters MONITOR. In MONITOR, the overall system current is first measured and subtracted from the system limit. Any time the measured current exceeds the system limit the program falls into FIND FIRST SHEDDABLE. This program starts at the top of the (APN) list and proceeds downward until the first device drawing current is found. It then obtains the control code and goes into SHED MODE SELECT. On the basis of the control code it will decide what to do about the load. Since we are at start-up and no loads have been turned on except the 00 loads, the device would fall through the decision tree and wind up at the extreme right giving the "00" alarm. This would indicate that the non-sheddable loads were exceeding the system limit and that something must be manually disconnected.

**7-3**   A control algorithm.

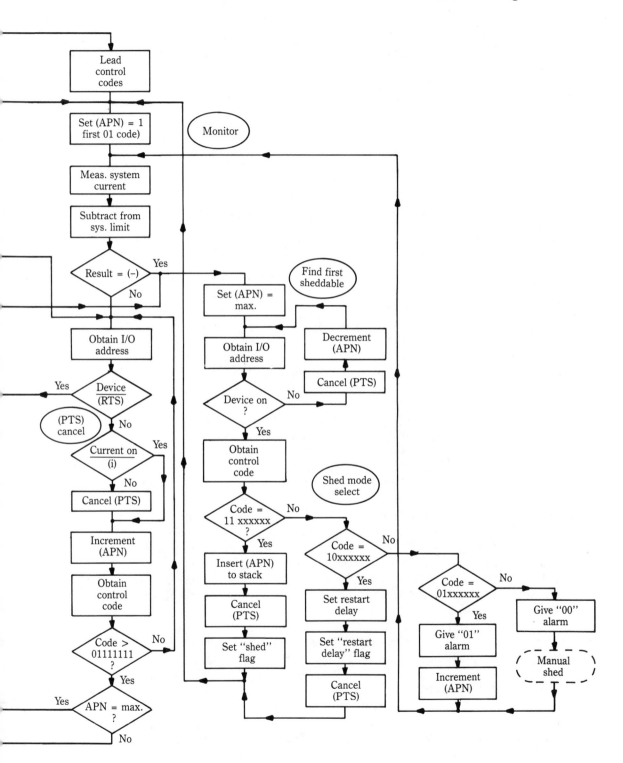

In the case where the system load was within tolerance, the processor would obtain the I/O address of the first 01 (deferrable but non-sheddable) device and determine whether it was requesting a start (RTS). It is rather likely that at start-up the device would be requesting a start since the system has been down. In this case, the control would proceed to the PERMISSION TO START routine.

In PERMISSION TO START, the processor would first obtain the control code and compare the current rating of the device with the margin between current being drawn and the system limit or capacity. If the system capacity would not be exceeded, the processor would set the start delay taken from the CONTROL CODE and would initiate the start. When the start delay has timed out, the control jumps back to the start of MONITOR, resetting the value of (APN).

On the second time through, the first device will no longer be requesting a start. Therefore, control will fall through (PTS) CANCEL. If the device was not drawing current, (PTS) would be canceled at this point. In any event, (APN) is incremented and the new control code compared with 01111111, which is the code for the last deferrable, "non-interruptible" load, and the next device will be tested for a start code after passing through MONITOR.

It can be seen that all the 01 devices will be added as fast as the starting delays will permit until the current exceeds the system limit or until the current code indicates that the system limit would be exceeded. In the case where the system current is not exceeded but the device code indicates that it would be, the system will bypass that device and will search for a device that could be started without exceeding the current limit. Since the search always starts at the bottom of the 01 list, the control will always be polling all these devices and will cancel (PTS) as soon as any device ceases to draw current. A high priority device will therefore be started as soon as doing so will not exceed the system current limit.

In actuality, this bypassing would involve a pass through FIND FIRST SHEDDABLE and fall through the SHED MODE SELECT tree. At start-up, this would involve issuance of a brief "01" alarm and an increment of the (APN). This loop will assure at later stages of the operation that whenever a high-priority load is denied permission to start, at least one of the active, lowest-priority loads is shed. Because (APN) is reset to the bottom whenever a load has actually been shed, the control will climb through the (APN) list to the device and try again. As long as any 11 or 10 devices are active, the control will continue to shed load and try again. At start-up, of course, there are no 11 or 10 loads to shed. Therefore, it is necessary to bypass the device. At startup, the devices will not be serviced on the basis of strict priority alone.

Only when the control can fall through (PTS) CANCEL without encountering any requests to start (RTS) and the system begins encountering 10 and 11 addresses, does the system test for 10 and 11 flags as the result of load shedding. At start-up, of course, there will be no such flags and the system will attempt to start all 10 and subsequently all 11 loads requesting a start. Eventually, if the system current limit is adequate, the control will fall through all (APN)s without a start request and the system will be in equilibrium. When this happens, the control will cycle through the MONITOR and (PTS) CANCEL loop waiting for the next (RTS). If the system load is running fairly light, a request from a 01 device will necessitate only the shedding of one or two 11 devices, and the system might never get into the 10 devices. In this case, the system would fall through RESTORE to the load SHED FLAG test. The second RESTORE loop is set up to find the load that was shed earliest, which will not exceed the current limit. The arrangement is intended to provide for the fastest possible return of any load that has been shed without regard to priority. If no SHED flag is set, the control polls the 10 and 11 loads in order of priority.

If a 10 device has been shed, the control decrements the delay register and tests to see whether the delay has timed out on the shed 10 devices. When the time-out is complete, the device is cleared for restart. It should be noted that this time-out is somewhat variable since any devices being started during the poll might have a start delay, which will slow the circuit of the loop. Since this makes the delay only slightly longer it does no harm to the system.

It will be noted that the operation of the RESTORE loop is such as to advance the priority of a previously shed 11 device ahead of all but the first 10 device and that would get precedence only if it had been shed earlier. If the system is operating reasonably within its capabilities, the shedding of a 10 load should be relatively infrequent and the servicing of previously shed 11 loads in precedence is probably proper. Too frequent shedding of 10 loads is a sign of system overloading. At this point either the use of some of the appliances should be discontinued or the system generation and storage facilities should be enlarged.

The algorithm shown here might not be optimum for all applications and might need to be enlarged or reduced to fit your requirements. Certain "bells and whistles" (i.e. day/night reprioritizing and active keyboard service to permit changes in parameters such as the system current limit without interrupting operation) have purposely been eliminated for the sake of simplifying an already complicated situation.

This algorithm should run fairly fast on most control-oriented processors. If no $(\overline{RTS})$ is encountered, the complete circuit should run in a few milliseconds. It might actually be necessary to place a few delays in the algorithm to permit the line a chance to stabilize. For control-oriented processors, the entire firmware can probably be coded in 1K bytes of ROM and the scratchpad memory requirements should not exceed 128 bytes if the algorithm is written at the machine language level. For high-level language machines such as the "appliance" computers now on the market, the requirement for memory would be considerably larger. Most of these machines would not have enough I/O lines to handle the various switches and sensors without an additional I/O board. Each 01,10, and 11 item will require three lines for $(\overline{RTS})$, (PRS), and $(\overline{9i})$. A total of 16 controlled items would require 48 I/O lines plus the lines needed for display, if any, of system current sensing and such.

# Power factor correction

The importance of having the power factor of the system high was mentioned earlier in the discussion of the motor-generator and generators in particular. We noted that the lagging current drawn by an inductor or an induction motor could cause the voltage of the alternator to sag, that a leading phase angle from a condenser could cause the voltage to soar, and that a proper power factor would maximize the capability of the system. In particular we showed in Fig. 1-4 that the correct capacitor would reduce the current drawn by the drill press motor from something over 5 amperes to something under $1/2$ ampere in the no-load condition. For the small home electrical system, a reduction from 585 VA to 56 watts is obviously worthwhile.

One might consider the possibility of correcting each motor individually for something like an average load condition since the power factor tends to change with load. This works reasonably well. However, it does not compensate for the starting surge that is much larger than the running reactive volt-amperes. We shall discuss a system that makes it possible dynamically to compensate the system power factor.

# The phase detector

The circuit of Fig. 7-4 is a sine wave phase detector similar to the circuit discussed with the phase-locked loop in Fig. 5-7. The transformer can be similar to the current transformer described in chapter 1. It can be seen that the magnitude of the voltage at A is the vector sum of the sample of the line voltage and the current sample that is in quadrature with the line current since the secondary of the transformer is very lightly loaded. The difference between the magnitudes of voltages A and B can be shown to be directly proportional to the amount of the reactive current being drawn by the load. The real parts of the line current cancel and the voltage samples are constant. Therefore, a dc analog-to-digital converter sensing the voltage difference between terminals A and B will provide a digital reading directly proportional to the item that we wish to cancel, namely the reactive amperes. Since we know the line voltage to be constant, our control computer could easily take this reading and calculate the amount of capacitance required to compensate the power factor.

# The capacitor bank

The illustration of Fig. 7-5 shows a suitable capacitor bank for power factor compensation in a home electrical system. You will note that the capacitors are arranged in binary sequence of values ranging from 20 $\mu$f to 640 $\mu$f. By selecting proper combinations in binary sequence the bank can draw anything from 0 to 55 amperes leading in 0.87 ampere steps.

In the figure, each capacitor is equipped with a double-throw switch going to a discharge resistor. If a relay or other mechanical switch is used to insert the capacitors, there is a fair probability that the capacitor will be disconnected from the line at a time when the capacitor is fully charged. If the switches are actually solid-state switches, it is a mathematical certainty that this will happen since the SCR or TRIAC will always turn off at zero current, which is precisely the time when the capacitor is fully charged. This is undesirable because the next switch-on might find the line voltage on the opposite swing and a large current surge would take

**7-4**  The phase detector.

**7-5**  The capacitor bank.

40 $\mu$F 330 Vac capacitors are GE type 21L3340
20 $\mu$F 330 Vac capacitors are GE type 21L3321

place in the first half-cycle. Accordingly, a second switch and a bleed resistor are provided to make sure the capacitor discharges rapidly after being removed from the circuit. If solid-state switching is employed, make sure that the switching is synchronized with the line voltage zero crossings to avoid transients.

It might be seen that the A/D converter is itself capable of generating the proper codes for insertion of the capacitors provided that a proper scaling is provided for voltages A and B. However, the use of the computer lends some further flexibility to the situation in that it can be used to anticipate the reactive surge of the motor start. Because of the requirement to obtain the magnitude of voltages A and B, there is a filter network in each. This network must have a delay of several cycles. This causes the correction to lag, and there is an initial large transient for a few cycles at each motor start. If the computer controller knows in advance that it is going to issue a start command to a given motor, it can easily look up the power factor correction required for that start and issue the power factor correction data to the capacitor bank simultaneously with the (PTS) issued to the motor. This can be used to minimize the starting transient.

# The capacitors

Note that the capacitors to be used in this application must be units rated for continuous duty across a power line of at least the voltage in use. High voltages and high temperatures greatly reduce the service life of the capacitors. The units noted on the drawing are rated at 330 volts. They are each a paper and oil type of capacitor in a sealed metal case. Current cost for this capacitor bank, if purchased individually, is approximately $500 without the switches or A/D converter.

# Economics

If we assume that the home electrifier has some considerable experience in electronics, the economics of the situation stack up about as follows:

| | |
|---|---:|
| Microprocessor kit | $ 200 |
| Input/output board | 150 |
| (RTS)/(PTS) & (i) units | |
| $45 each times 10 devices | 450 |
| A/D converters 6 bit, 2 required | 50 |
| Miscellaneous cases, devices, etc. | 200 |
| Power factor capacitors | 500 |
| Capacitor switches | 150 |
| Total | $1700 |

This cost assumes that the electrifier will work on the project without charge. At the present reading commercial load controllers start at $3000, and there are no available power factor correctors since there is no market for these devices among people who buy home electricity from the utility company with no penalty for poor power factor.

For the home electrifier the choice lies between relatively sophisticated and expensive load control and power factor correction on the one hand and the simple brute force approach of making the system large and robust enough to withstand the load transients on the other hand. Our "typical" average household usage of 1.16 kW can probably be handled by machinery capable of 3 to 4 kW with the sophisticated control and minimum restriction of service or by a 15 to 20 kVA system with little if any control of loading and power factor. Depending upon the nature of the system and the requirement for energy storage, not all the machinery might have to be enlarged. However, the difference in motor-generator size alone is probably a penalty in excess of the $1700 figure.

# 8

# The windmill

THE INVENTION OF THE WINDMILL IS LOST IN ANTIQUITY, AND PERHAPS the windmill is one of those things that have been invented a large number of times in a large number of places. Windmills of various sizes have spun in places as diverse as ancient Persia, Tibet, Portugal, and Holland. It seems likely that the Tibetan Prayer Wheel should be accorded the honors for the oldest design. However, a great many ancient types still revolve in various parts of the world. Some of these types are of interest to us here because of their simplicity. They can be built with the simplest of tools and materials. We shall examine a number of types to see how well they are suited to satisfying the appetite modern man has developed for energy.

After the development of agriculture, there was a desire to grind grain in order to remove the starchy edible portion from the indigestible cellulose husk. This can be done with a mortar and pestle by hand but it is arduous work. As early as the time of Samson (*Judges* 16:21), we find mention of a rotary grist mill with slaves plodding in a circle to rotate the millstones. It is little wonder that men sought cheaper and more efficient means of performing this work. It is also very logical that the machines developed in landlocked areas should be machines of different form than those developed in maritime lands where the use of the sail to extract energy from the wind was well known. The basic difference between these machines is in the direction of the axis of rotation. The landlocked machinery tends mostly to rotate on a vertical axis like a merry-go-round whereas machines of the sailing category generally rotate on a horizontal axis like an airplane propeller. For the reader interested in the history and lore of the windmill I recommend Volta Torrey's *The Wind Catchers,* published by The Stephen Green Press of Brattleboro, VT.

Substantial differences evolved in the design of the windmill because of the eventual use of the mill. When the object was to grind grain or to pump water from a deep well, it was naturally desirable to have the mill turn relatively slowly to minimize or eliminate the requirement for gearing. The production of a sturdy and long-lived gear from wood is no simple task and requires a great deal of both knowledge and craftsmanship.

The industrial revolution in the United States brought a change in the market for windmills. While the steam engine began to take over the requirement for large power in commercial enterprises, the expansion of the American West brought a large market for windmills to pump water and grind grain on isolated farms and homesteads. This in turn launched a very substantial industry, which at its peak in 1910, employed about 20,000 people in the US alone. There were a tremendous number of patents issued for improved wind engines, and a great many home experimenters built their own machines. After the end of World War I, the windmill industry declined. The interest in windmills fell to a low ebb except for a brief spurt in the 1930s when farmers became interested in a new form of electricity-generating machines that would power a radio on a nonelectrified farm. These were a different breed of machine, which turned much faster because of the high-speed requirements of the electrical generator. As Torrey mentions in his book, the construction of an electric wind engine is as different from a grain-grinding machine as a speedboat is from a barge.

# The wind

To begin with, let's consider a few of the known properties of the wind to see what effect they will have on the design of a usable windmill. The US Weather Bureau has spent a great deal of effort in recording and analyzing wind patterns and velocities, and these can tell us a good deal about what we can expect to achieve and what we might have to put up with in a wind-driven engine.

Experience indicates that there are well-defined groups of winds that predominate and can be called the *prevalent winds*. There is a second group that contain the bulk of the energy; these might be called *energy winds*. The reason for the distinction is that the energy present in a column of wind in front of your windmill is proportional to the cube of the velocity. This is a very rapidly rising function as can be seen from Table 8-1. I have normalized the table to 15 mph for reasons to be discussed later.

From the table you can see that an increase in wind velocity from 15 mph to 20 mph more than doubles the available energy, and a decrease to 10 mph nearly quarters it. Weather bureau measurements indicate that energy winds, even in calm summer months, produce about 70 percent of the energy even though they blow only about 42 percent of the time.

Furthermore, the following facts obtain: Energy winds blow 2 out of 7 days and prevalent winds blow 5 out of 7 days. The mean prevalent wind velocity is 2 mph less than the average monthly velocity. The wind of the highest energy is about 10 mph faster than the most frequent wind. The total energy per month is about twice the energy that would be computed using the average monthly velocity. If the windiest and the calmest months are compared, the energy for the windi-

**Table 8-1. Relative energy
in relation to wind velocity.**

| Wind velocity | Relative energy |
|---|---|
| 5 mph | 0.037 |
| 10 | 0.296 |
| 15 | 1.0 |
| 20 | 2.37 |
| 25 | 4.63 |
| 30 | 8.0 |
| 35 | 12.7 |
| 40 | 18.96 |
| 50 | 37.0 |
| 75 | 125.0 |
| 100 | 296.3 |
| 125 mph | 578.7 |

est will be about 4.5 times as great and the wind velocity 1.65 times as great. The true velocity will depart from the average velocity by $\pm 28$ percent in a 10-second period and $\pm 45$ percent in a 38-second period. The high-wind regions of the US (having 10 mph or more average yearly velocity) are a 350-mile strip north and south through the center of the nation, the Great Lakes drainage basin, the Atlantic Seaboard, and the Pacific Coast near San Francisco and Washington. The fastest recorded wind in an area is likely to be about 9 to 10 times faster than the yearly average.

Over unobstructed terrain, the wind varies with height with the relationship:

$$\frac{V}{V_o} = \left(\frac{H}{H_o}\right)^n$$

where $n = 1/5$ from 5 to 35 mph.

A telephone call to your local weather bureau will probably yield a yearly average wind velocity at the weather station and factors such as the highest recorded wind. However, the first logical step for anyone seriously contemplating the erection of a wind engine is to put up an anemometer and begin the rather lengthy task of accumulating data for the specific location. As we shall see shortly, the design of the machine you intend to use can be considerably influenced by the wind velocities expected in the area.

# Windmill requirements

Once written, the following rules are likely to seem self-evident. However, neglecting to observe these rules is responsible for the overwhelming majority of windmill failures, which can result in property damage and injury as well as loss of investment. To be effective and safe a windmill should be:

- Equipped with a brake to permit stopping of the rotor
- Designed to operate in the expected energy winds
- Equipped to respond to shifts in the wind
- Designed to withstand gusts in operation
- Designed to survive maximum wind velocity, without blowing down or disintegrating from centrifugal force because of over-speed

On all but the smallest windmills, the failure to observe these precautions in the design of the windmill will almost certainly result in disaster.

# The vertical axis machine

The vertical axis Persian kind of windmill is thought by many to be one of the oldest designs since it is so elementary. As shown in Fig. 8-1 in the top view, the rotor consists of four or eight flat paddles mounted on a vertical shaft. A small building is constructed to shelter a little more than half of the wheel from the prevailing wind. The structure also serves to provide a top bearing support. Obviously, with a machine of this sort, the tip speed cannot exceed the wind velocity. In fact, it will usually run more slowly because of the air being churned by the unused half. It is also fairly obvious that the machine is of use only in areas where the strong winds are fairly fixed in direction. When it is necessary for lubrication or service, the machine can be stopped by closing the opening for the wind.

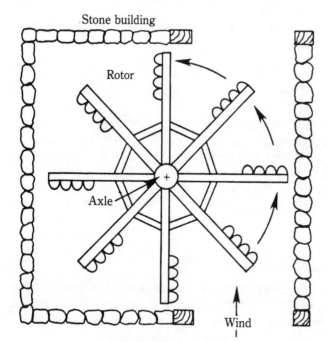

**8-1**  The Persian windmill.

A more modern version of this type of machine is found in the anemometer shown in Fig. 8-2. In this machine a set of hemispherical cups is arranged around the periphery of a circle. The open cup facing toward the wind has about four times the drag that the cup facing closed-side toward the wind presents. Therefore, the entire assembly might be exposed to the wind. This arrangement is commonly used for the measurement of wind speed since it does not need to be faced or aimed into the wind. Here again the tip speed is something less than the wind velocity.

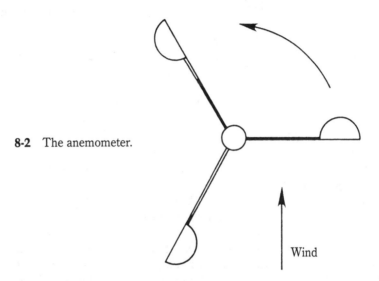

**8-2**   The anemometer.

Wind

The Savonius windmill shown in Fig. 8-3 is a member of this same family. This can be made 2- or 4-bladed. In the 2-bladed format, it is felt that the wind spilling along the flat side increases the efficiency. In the 4-bladed configuration, it is more usual to leave the central portion as an open frame. A popular American version of this form was constructed using 55-gallon oil drums cut in half parallel to the axis as the cups. This provided a neat-but-not-gaudy homebuilt design.

For machines with cylindrical cups, the air drag with the open face toward the wind is about 1.3 times the drag of a flat plate of the same area, whereas the drag of the cup with the closed face toward the wind is about a third of the drag of a flat plate. The net static torque is therefore about equal to the torque that would be experienced by a flat plate with the same area as the radius of the cup. The pressure on a flat plate due to wind is given as:

$$P = 0.0042 \ V_r^2$$

where $P$ is in pounds per square foot and $V_r$ is the velocity of the wind relative to the surface in miles per hour. Taking our oil drum example, let us suppose that the area of the open face is 2.5×4 feet or 10 square feet. In a 30-mph wind, this would experience a force of 37.8 pounds, with the rotor stalled or locked. If the rotor had a radius of 8 feet, the stalled torque would be 302.4 foot-pounds. This is a tremen-

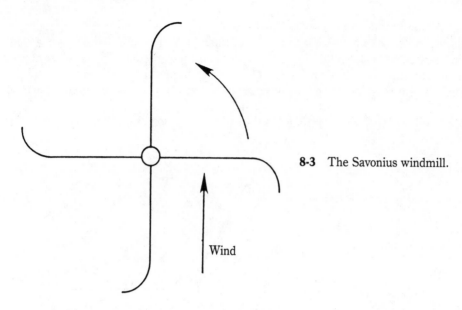

Wind

**8-3** The Savonius windmill.

dous amount of torque, but we are not yet extracting any work from the machine because it is not turning.

When we consider the turning of the machine, we have to consider several other things. First of all, once the machine starts to turn, the working cup (open mouth) is going downwind and the relative speed of the wind decreases by its speed. On the other hand, the other cup is traveling upwind and the relative velocity it sees is increased by the rotational velocity. If we consider the wheel to be turning so that the cups are traveling 10 mph for the working cup we would obtain:

$$F = 4/3 \times 0.0042 \times (30-10)^2 \times 10 = 22.4 \text{ lb}$$

and for the dragging cup we obtain:

$$F = 1/3 \times 0.0042 \times (30 + 10)^2 \times 10 = 22.4 \text{ lb}$$

Since these forces are opposed to one another, it says that this windmill would be yielding zero net torque at 10 mph rotational speed. The more mathematically inclined reader will see that the speed of the mill is firmly fixed by the square root of the ratio of the forward to the reverse drag of the cups or:

$$\frac{V\text{max}}{V \text{ wind}} = \frac{\dfrac{(D_f)^{1/2}}{(D_r)} - 1}{\dfrac{(D_f)^{1/2}}{(D_r)} + 1}$$

At a cup speed of 5 mph, the working cup will be developing a force of 35 lb and the dragging cup a force of 17.15 lb for a net difference of 17.85 lb. The machine would have a torque (neglecting friction) of 142.8 foot-pounds and would be developing a power of 0.238 horsepower. It would be turning at 8.75 rpm. You can see that this is not exactly the ideal machine for generating electricity when we would like to have our alternator spin at 3600 rpm or more for an automotive alternator/rectifier set. The machine is well suited for heavy pumping and grinding where the tremendous static torque would be of value.

One is inclined to question whether there is not some way in which the tremendous static torque of the machine could be preserved while the $D_f/D_r$ speed limitation was defeated. One such proposition would be the epicyclic machine shown in Fig. 8-4. Each cup is held by a jointed arm and pivots on the end of the arm. The arrangement permits the dragging cup to travel upwind at a very small fraction of the downwind speed of the working cup, thereby minimizing the drag loss and also the drag torque. The balancing and construction of this machine would be no trivial engineering feat, and the machine still could not run as fast as the wind. However, it would be so wondrous to behold in operation that it seems a shame that someone does not take the effort to build one!

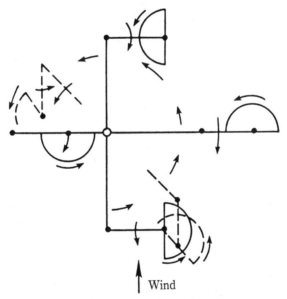

Wind

**8-4**   The wonderful epicyclic vertical shaft windmill.

There is another class of vertical shaft machines referred to as the Darrieus machine, which resembles a giant eggbeater. These machines can operate faster than the wind and do not require steering. They will be discussed later under the heading of aerodynamic machines. These simpler machines are essentially aerostatic in nature and drift with the wind in much the same way that the square-set sail on an ancient galley did.

# Aerodynamics

The science of aerodynamics is extensively based upon the principle announced by Daniel Bernoulli in 1745. In essence, Bernoulli's principle states that the total energy in a moving column of a fluid is a constant. One of the sequelae of this is the fact that if some local disturbance increases the velocity, thereby increasing the kinetic energy, the pressure must fall. This in turn decreases the potential energy so that the energy in the column can remain constant.

You can convince yourself of the operation of Bernoulli's principle very easily with the simple experiment shown in Fig. 8-5. A funnel is taped onto the end of the *blowing* hose from a household vacuum cleaner. If a ping-pong ball is then inserted in the funnel, you will find not only that the air does *not* blow the ball out, but that the funnel can be inverted and the ball will not even fall out! The explanation is that the air rushing through the gap between the ball and the funnel attains a lower pressure than the atmosphere. The atmospheric pressure tends to hold the relatively light ball down in the throat of the funnel.

**8-5** A demonstration of Bernoulli's principle.

As a further refinement, a water manometer tapped into the funnel near the point where the ball is closest to the funnel will show a pressure difference of one or two millimeters of water. The ball will float in the funnel with some oscillation and will perhaps rotate on some random axis while maintaining a small annular gap with the funnel.

The operation of Bernouilli's principle is shown somewhat schematically in the illustrations of Fig. 8-6. If we consider that the wing is stationary and the air is flowing from left to right, we observe in the top illustration that the air on the top of the wing takes a more curved and therefore longer path to get from the front to the rear of the wing. It follows that the velocity must be higher and its pressure lower than the air on the bottom of the wing.

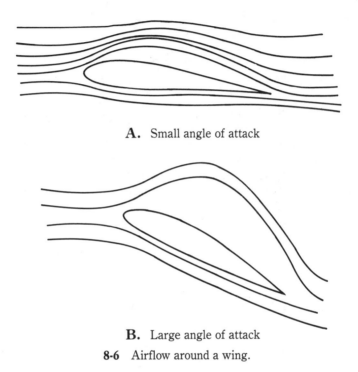

**A.**  Small angle of attack

**B.**  Large angle of attack

**8-6**  Airflow around a wing.

In the lower illustration, we observe that increasing the tilt of the airfoil with respect to the airstream will increase the lift (up to a point) by making the disparity between the upper and lower paths greater. Patterns such as these might actually be seen in a wind tunnel by releasing streams of smoke from a pipe with regularly spaced holes. The streams make the airflow visible.

Beyond a certain limiting angle, the flow of air above the surface of the wing ceases to be smooth and orderly or "laminar" and becomes disordered or "turbulent." At this point, the wing loses most or all of its lift and the airfoil is said to have "stalled." Figure 8-7 illustrates this situation. This property used to be widely employed in landing airplanes equipped with a tail wheel. The pilot would bring the airplane down so that its wheels were only a foot or so above the runway with no engine power. As the airplane lost speed because of drag he would gradually bring the nose up further and further to increase the lift enough to just hold the altitude constant. With increasing angle of attack the drag would also increase and the airplane would slow further. At some speed, the wing would suddenly stall and

**8-7** The stalled wing.

the airplane would gently drop the two or three inches for a perfect "three-point" landing, coming down with the minimum forward velocity possible for the plane.

The pressure distribution for a typical airfoil operating at a moderate angle of attack is shown in Fig. 8-8. Note that the vertical scale is reversed, reading minus in the upward direction. This is usually done to keep the low-pressure readings in association with the top of the wing and the high-pressure readings associated with the bottom of the wing.

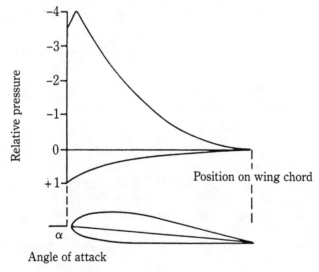

**8-8** Pressure distribution on the wing.

Figure 8-9 shows some of the parameters of the airfoil including the definitions of the airfoil thickness $t$, the chord, and the angle of attack. The differences in pressures between the two sides of the airfoil are balance about some point called the *center of pressure*, *CP*. This pressure is actually normal to the chord. However, a force is also attempting to resist the motion of the airfoil through the air. Therefore, the net force on the airfoil leans back somewhat from the normal *N*. The

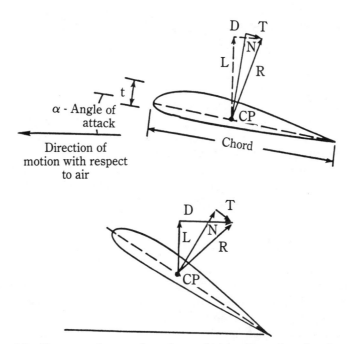

**8-9**  Forces acting on the wing and the effect of angle of attack.

resultant force is labeled $R$ in the figure and a retarding force labeled $T$ can be inferred by resolving $N$ and $R$. Actually we are more interested in the forces at right angles to the motion of the airfoil designated *lift* or $L$ and the force opposing the motion of the wing in the direction of its travel, which is called *drag* $D$. From the lower illustration we see that a tilt of the angle of attack also tends to rotate $R$ through the same angle. On a real airfoil $R$ also changes its angle with respect to the chord. However, the most significant feature is the fact that $D$ increases dramatically. In essence, we have rotated some of our lift into drag by tilting the wing. In actual practice, lift actually increases up to some point, but drag increases faster and the $L/D$ ratio is generally greatest at some small angle of attack for thick airfoils.

Both the lift and drag coefficients of an airfoil are a function of the square of the velocity of the airfoil with respect to the air. This speed determines the amount of load that an airfoil can handily lift. This is usually given as wing loading in lb/square foot. Table 8-2 shows some typical working numbers.

**Table 8-2. Speed determines amount of load.**

| Aircraft | Speed | Wing loading |
|---|---|---|
| Gossomer Condor | 8 mph | 0.15 lb/ft$^2$ |
| Piper Cub | 75 | 12.5 |
| Curtis P-40 | 225 | 120 |

This works out to a working wind loading of:

$$WL = \left(\frac{V}{21}\right)^2$$

where $WL$ is in lbs/ square foot and $V$ is in miles per hour.

Note that this is only a rough-and-ready rule of thumb and that it included a variety of different airfoil cross sections. In general, the slower an airfoil is intended to operate the thicker the section should be with respect to the chord, or at least the greater the curvature of the upper section should be. The wing sections on the Piper Cub and the P-40 tend to run a maximum thickness of about 14 percent of the chord. The selection of aircraft was made to include the major range of speeds liable to be encountered in a windmill design.

# Relative speed

The factor that tends to make the aerodynamic windmill attractive compared to some of the aerostatic types mentioned earlier is the ability of the device to operate at speeds higher than that of the wind. To understand this, we must examine what actually happens from the viewpoint of some piece of the blade of a large propeller.

Figure 8-10 shows a section of a windmill propeller. The velocity of the wind is represented as an arrow and the rotational velocity of the blade is represented as a second arrow. Because the blade is approaching a given element of air from right

**8-10**   The apparent wind on a blade section.

angles, and the element of air is also approaching, the two velocities add in the triangular form shown and the wind appears to the blade to be approaching at some small angle of attack. The resultant thrust on the airfoil is tilted slightly backward from the direction of travel. However, it still has a significant component in the direction the propeller blade is moving mechanically. We designate this force $MT$ for *motor torque*. There is also a significant component or $R$ *crosswise to MT*, which we designate $EL$ to stand for the *end loading* that must be born by the thrust bearings on the shaft and the tower itself.

The component $MT$ is the portion of the work that can be presented to the windmill shaft to do useful work. Now let us suppose that for some reason the resistance of the output work were to decrease. In this case, $MT$ would accelerate the propeller and increase $V_{blade}$. In this case, it can be seen that the wind triangle would flatten and $V_{apparent}$ would swing around slightly so that the angle of attack, $\alpha$, decreases. Under a no-load condition the speed would actually increase until ($\alpha$) became very small or even went negative and $R$ would shrink to an equilibrium level, just large enough to make up friction losses.

On the other side of the coin, if the load suddenly increased, this would slow the blade slightly, thereby increasing the effective value of ($\alpha$) and increasing the magnitude of $R$. The component $MT$ would then rise, and the machine would find a new equilibrium delivering more torque at a slightly slower speed.

If the torque load were to slow the blade enough, the angle $\alpha$ would exceed the stall angle for the blade and the machine would quickly stop. This stall property is more severe on very high-performance machines.

# Helical pitch

Having seen that there is a critical relationship between the angle of the blade and the speed of the blade with respect to wind speed, we note that the speed of the blade segment is a function of the radius to that particular segment. A segment that was twice as far out would have twice the blade speed and one that was half as far out would have only half the blade speed. In order for each to operate with reasonable efficiency, each would have to be adjusted to a different mechanical angle with respect to the shaft in order to face the apparent wind properly.

Figure 8-11 shows this. The dimension labeled *pitch* is the equivalent of our wind velocity. This is the distance that each segment of the propeller would advance in one revolution through the air without any slippage. The design is shown for a propeller whose tips would rotate at a theoretical speed equal to six

Position along blade                      Center
Periphery

**8-11** The requirement for helical pitch.

times the velocity of the wind. It can be seen that the outer portions of the propeller are very flat compared to the central portions.

Another point is noteworthy. The innermost portion shown is traveling slightly less than a quarter of the speed of the outer section shown, and therefore for equal area, would give only about 1/18.5 as much lift. The central portions of a propeller yield only a small portion of the motor torque. The requirement for a helical pitch and the relative ineffectiveness of the central portions of the windmill were recognized very early by some millwrights in Holland. However, other countries did not necessarily observe this phenomenon. In the typical Dutch windmill, the central fifth of the wheel is empty and very pronounced twist in the blades is noticeable. French windmills with perfectly flat blades were relatively common. While a mill will work with flat blades, the efficiency is much lower, and the machine will deliver considerably less torque at any given rotation speed since some portion of the blade is always stalled.

As one goes further and further out on the propeller, the flatter angle of approach to the wind makes the achievement of a very good lift/drag ratio more and more important. This is shown in Fig. 8-12. Here the blade is shown with a velocity six times the velocity of the wind. Compared with the section shown in Fig. 8-10, we would expect the total force $R$ on the blade section to be:

$$\frac{F_{\text{outer}}}{F_{\text{inner}}} = \left(\frac{6.08}{1.41}\right)^3 = 80.29$$

**8-12** The lift-to-drag ratio effects on outer blade portions.

However, you will note that unless the drag tilt is less than 9.46°, there will be no net contribution to the value of $MT$. Stated differently, an airfoil section cannot contribute to the net torque unless the lift/drag ratio exceeds the ratio of apparent velocity/wind velocity. This implies several factors:

- For very high speeds with respect to the wind, the airfoil must be a basically low-drag type.
- The airfoil must be operating near the angle of attack that yields maximum $L/D$.
- The airfoil must be very smooth to minimize parasitic drag.

While certain airfoils can yield lift/drag ratios in excess of 30, the state-of-the-art in propeller design seems to limit maximum no-load speeds for windmills to about 11 times wind velocity and maximum power is usually obtained at about 6 times wind velocity.

# Tip vortex

At the tip of the blade, a peculiar condition occurs in that there is a low pressure on the one side of the blade and a high pressure on the other. This gives rise to a tendency for the high-pressure air to wash around the end of the blade to create a vor-

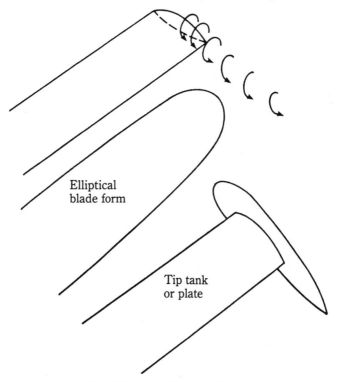

Elliptical blade form

Tip tank or plate

**8-13**   Blade tip effects and vortex.

tex that trails the tip. This adds to the drag and tends to spoil the *L/D* ratio of at least the outer portion of the blade. This effect can be minimized by:

- The addition of tip plates or tip tanks
- Design of the blade to have an elliptical cross section
- The use of very large diameter to blade-width ratios

Figure 8-13 shows the phenomenon along with several of the proposed solutions.

At relatively high speeds, the addition of judiciously designed tip plates or tip tanks can considerably reduce the drag. It was discovered during WW II that the addition of tip tanks to fighter planes could often raise the cruising speed by 4 or 5 mph. On the propeller, however, the addition of the mass of the tip plate or tank increases the problems of centrifugal force. For many years the wings and propellers of high performance aircraft were designed to present an elliptical form to minimize this effect.

For high-efficiency soaring planes, the wings were simply built with a very high aspect ratio. Gliders with very long narrow tapering wings seem to perform much better than those with short stubby wings because of the tip vortex effect. In essence, since the vortex affects only the region near the tip, this is made the smallest possible percentage of the wing.

On the basis of this very brief discussion of aerodynamics, we shall next discuss the expected performance of a variety of windmills in the sail and propeller classes.

# 9
# Windmill types and performance

IN THIS CHAPTER WE SHALL EXAMINE A FEW OF THE WINDMILL TYPES with an eye to exploring their relative advantages, disadvantages, and limitations. We shall proceed in roughly chronological order.

## The Portuguese or jib mill

Figure 9-1 shows the Portuguese or jib mill in front and side views. In general, these types, which today may be seen in operation in Portugal and on the island of Crete, consist of a series of six masts mounted radially about the windshaft and carrying six jiblike sails. The sail is very similar to the jib of a small sailboat and is sheeted to the tip of the adjacent mast. Two sets of stays are used; one set ties the tips of the masts together, and the second braces the mast against the force of the wind from the shaft, which juts forward of the masts. At the hub, opposing pairs of masts are mortised to take the windshaft and lashed together. This provides a crude but sturdy mechanism for transmitting the torque moment from the masts to the windshaft.

As can be seen from the illustration, the chief advantage of this type of mill stems from its very crudity. It can be built by nearly any competent carpenter with a saw, chisel, hammer, and an adequate supply of wood, cloth, and rope.

This mill also has the advantage that it can easily be made large enough to produce significant amounts of power in very light winds. It turns slowly with the tip speed seldom exceeding the wind speed because of the relatively poor airfoil shape blown into the sails.

The chief disadvantage of this type of mill is that the sails require rather precise adjustment to compensate for changes in the wind. Much like a sailboat, on changing wind the jibs require more or less continuous attention, either taking up the sheets or paying them out as the wind force changes. To alert the miller to the

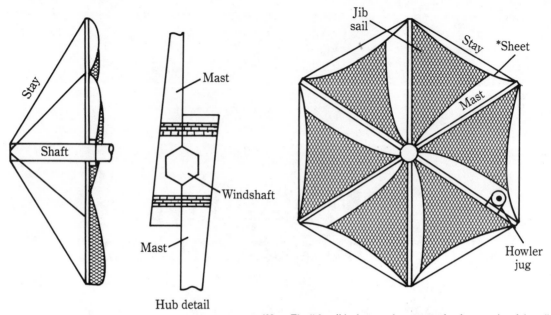

Hub detail

*Note: The "sheet" is the rope that secures the clew or point of the sail.

**9-1**  The Portuguese or Jib mill.

situation, the masts were frequently festooned with large-mouthed jugs facing the wind. As the tone of the howling of the jugs changes, the miller is alerted to the requirement to brake the wheel to a halt, go out and change the bottom sheet, allow the wheel to turn a sixth of a turn, and repeat the process until all six sails are reset. In higher winds, the sails are reefed progressively.

It is possible for a mill of this type to be built using roller-reefing gear and some type of roller-sheeting gear operating through a hollow windshaft; however, this takes the machine out of the class of simple machines. With such attachments, however, the machine could be expected to achieve tip speeds equal to or exceeding wind velocity. For light-air locations, there is probably no cheaper wind engine.

# The Dutch windmill

Figure 9-2 is intended to depict a basic compendium of the Dutch windmill. It is really not very fair to summarize the wide variety of exquisitely designed and crafted machines developed in Holland over a number of centuries with a single drawing and a single figure. However, the drawing does show at least some of the cardinal features of these machines. In general, the opposing masts were square and had the same butt thickness as the square windshaft. They were generally spliced with a pair of timbers that also captured the windshaft. The second set of masts ran behind the first.

The masts were tapered and carried a set of relatively light battens and stringers, which supported the sail. The views showing the inner and outer sections

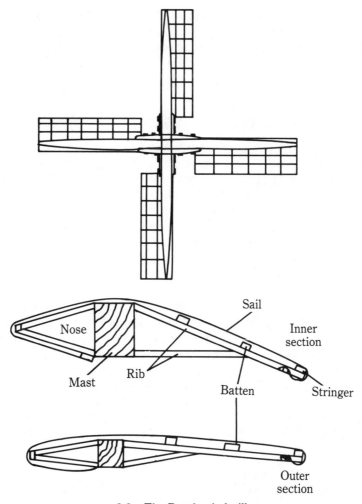

**9-2** The Dutch windmill.

illustrate the means by which the change in airfoil thickness and angle of attack were obtained.

On many of the later mills, the sailcloths were replaced by a series of spring-loaded, hinged shutters attached to the ribs. This permitted the shutters to swing open during sudden wind gusts to protect the mill from overspeeding. In some later versions, it also removed the requirement to stop the mill in four different positions so that the miller could furl the sails.

The Dutch windmill would attain no-load tip speeds of four times wind velocity and maximum power at about two times wind velocity. The medium width-to-diameter ratio on the order of six contributed significant tip vortex losses. The open rib framework and general aerodynamic clutter of the ribs, mast, and battens tended to limit the speed and therefore the output of the machines. However, given the state of the knowledge of aerodynamics at the time when these machines were built, the remarkable feature is the use of helical pitch, the tapering airfoil section, and the achieved cleanliness and performance. Apparently without any formal science of aerodynamics, the Dutch millwrights learned by trial and error most of the basic parameters of windmill design.

# The American wind turbine

From the very earliest settling of the United States, there had been a large number of windmills of the Dutch type used for grain grinding and other industrial purposes. However, with the beginning of the settlement of the plains states there came a demand for a new type of wind machine to pump water from the deep wells required in the dry plains. More significantly, a new requirement arose in that the machines had to be capable of unattended operation. Whereas the miller or the blacksmith would be at hand to swing the windmill into the wind or furl the sails when the wind got too high, the farmer would like to have his stock watered even when he was in a distant field and not available to care for the windmill. This led to nearly a century of development of the unattended farm windmill. Although American inventors managed to produce a bewildering variety of these machines, there were two basic configurations that eventually came to dominate. The *iron turbine* shown at the left of Fig. 9-3 usually had 12 to 18 blades of sheet iron curved into an airfoil section. The blades could be either overlapping or nonoverlapping when viewed from the front. Because the high-pressure stream from the bottom of one blade passed over the low-pressure stream from the adjacent blade, the stalling of the blade was significantly retarded, and the machines would develop tremendous torque at very low speeds. This is a valuable property for a machine intended to handle the alternating load from a reciprocating well pump. When the wind

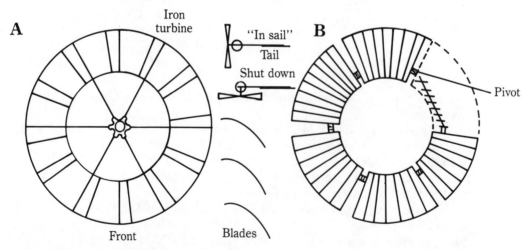

**9-3**   The American wind turbine. The shutter type with one segment shown blown down "out of sail" is at B.

became too vigorous, the torque developed by the wheel would swing the wheel out of the wind, parallel to the flat rudder or tail that normally keeps it pointed into the wind.

The *shutter* type had a large number of segments of very much smaller blades placed closer together. The segments were usually spaced with a few "missing teeth" for clearance. In this type, the segments would blow down roughly parallel with the wind when the wind became too strong. The blades or entire blade fans on these units were often made of light wood; therefore, fewer of these survive. The shutter type had an advantage in lighter air because of the lightness of the blades.

The iron turbine was capable of attaining a tip speed of something like 1.75 times wind speed and yielded maximum torque at something like 3/4 wind speed. Windmills of this type are still available commercially from a few sources that sell them primarily for water pumping.

## Power extraction

All windmills obtain power from the wind by slightly slowing the wind going through the wheel. Obviously, the wind cannot be brought to a dead stop behind the wheel since this would cause a "pileup" of air behind the wheel, which would stop the propeller. It can be shown mathematically that no windmill can ever extract more than 59.2 percent of the power from the air column. Windmills are therefore rated on the basis of power coefficient ($P_c$), which is the fraction of the available power extracted. The curve of Fig. 9-4 shows the comparison between the efficiency of the various types. It should be noted that the American wind turbine is a relatively modern machine of comparative aerodynamic sophistication compared to the Dutch type represented. With comparable aerodynamic cleanliness, the Dutch machine would rise vertically and to the right.

The unmistakable conclusion from the data, however, is the fact that the nod for overall efficiency must go to the high-speed, two-blade type of machine. In the recent surge of work in energy-related areas, it is only the high-speed machines of the propeller and Darrieus types that have received significant attention.

## How much power can you get?

The answer to this question gets a bit more sticky. In general, the power derived from a wind engine is a function of the aerodynamic cleanliness, the tip speed/ wind speed ratio for maximum power, the square of the radius of the rotor, and the cube of wind velocity. Unfortunately, the reported powers from windmills are nearly always given at the highest wind encountered when the windmill was tested. Depending upon the location, this might have been a 10- or 35-mph wind. In one case, a 70-mph wind was used for the machine rating. The curve of Fig. 9-5 was made up from data from various sources. The data have all been normalized by the cube-of-wind-velocity ratio to a 15-mph wind velocity, which would correspond to an energy wind for about 75 percent of the country. The curve includes

PC

High-speed two-blade
propeller

American wind turbine

Dutch 4 blade

Tip speed/Wind speed

**9-4**   Relative power extraction coefficients from wind tunnel tests.

some relatively small machines along with the largest and most powerful windmill ever built. Table 8-1 may be used to extrapolate the data to other wind velocities.

Going back to our energy wind concept, if we presume that 70 percent of our energy would be produced by energy winds blowing 42 percent of the time and we take our energy wind as 15 mph, we may look at the 2-kW point and see that a 30-foot *very efficient* windmill would give us:

$$2000 \text{ W} \times 1/0.7 \times 0.42 \times 24 \text{ hrs/day} \times 30.44 \text{ days/month}$$
$$= 876,600 \text{ watt hr/month}$$

or an average power drain of 1200 watts. If this energy is twice the energy for a wind of average velocity, the average wind velocity would be 13 mph, which is available only in the high-wind areas of the country.

It is also noteworthy that if energy winds blow only about 3 out of 7 days, we have to be prepared either to store rather considerable amounts of energy or to content ourselves to operate the high-drain appliances only at times when the energy winds are blowing. The 1.2 kW figure does not track too badly with our average household drain computed in chapter 1 if we presume that the home electrifier will have developed some sense of conservation by the time he sees how much effort is required to extract power from the wind.

The strength and frequency of the winds is obviously a very central issue in the selection of windmill types. It is generally considered that an 8-mph yearly average represents the minimum wind velocity practical for a 2-blade propeller type. Below this speed some multiblade type has the clear advantage. It has been shown that a very light multiblade wheel, which must be swung out of the wind at 15 mph and operates from 6 to 15 mph, will produce a monthly average power 14 percent greater than a heavier engine that operates only from 8 to 15 mph.

A small multiblade-type windmill, similar to the farm varieties used for water pumping, is generally not capable of driving a modern high-power automotive alternator to full output in winds less than 20 mph as seen in the curve of Fig. 9-5, combined with the multiplier of 2.37 from Table 8-1. In the light-wind areas that make up a significant fraction of the country, a very large windmill is required in order to come up to the 1-kilowatt monthly average. Something on the order of a 50-foot Portuguese wheel with reefing and sheeting handled by electric motors under automatic control would probably do the job, but the expense for a structure of this size would be very large and the design of a safe, stable, and efficient structure would be a major assignment in mechanical engineering. It is highly questionable whether the development and construction of a wind-electric plant for the home can be justified for an area where the mean-annual wind is less than 8 mph.

The curve of Fig. 9-6 gives the propeller diameter required for a two-blade propeller-type windmill producing an annual average of 1 kW, assuming that the power derived is twice the power available in the mean wind. The range of the curve covers the least windy areas where a propeller engine is practical to the windiest area in the United States, Mount Washington. The extreme sensitivity of the curve to wind speed is apparent. In a typical "windy" area such as Albany, NY (9.0 mph), the propeller is a 25-foot giant. In Buffalo, NY (14.6 mph) it has shrunk to 12 feet, and atop Mt. Washington (36.0 mph) it is a 3-foot pigmy!

It should be noted that 39-mph winds are not unusual for Albany. At these times the machine would be producing 3.4 kW. However, days when the wind is less than 5 mph and when the windmill would hang stationary are not unusual either. If the energy wind is assumed to blow 42 percent of the time and produce 70 percent of the power, the energy wind works out to 39 mph. At the same time the maximum recorded wind in Albany is 71 mph. Good engineering practice would probably call for design of the structure to withstand 100 or 120 mph winds.

# Forces of the windmill

The diagram of Fig. 9-7 illustrates some of the principal forces acting upon the windmill. If we consider a small segment of the blade, we see that it is acted upon by the motor torque ($MT$) and the end-loading force ($EL$), both of which are due to the aerodynamic performance of the blade. It is also acted upon by centrifugal force due to its own weight, velocity, and radius of rotation and it must sustain the action of centrifugal force due to the portions of the blade outside of itself. By tilting the blade slightly backward, the centrifugal force can be made to counteract

**9-5** Power out versus rotor diameter (normalized to 15 mph average wind).

$$kW = 9.17 \times 10^{-8} \ D^2 \ V^3$$

D = Diam.-ft
V = Velocity-feet/sec

(Note: This formulas has been de-rated from the usual theoretical formula to track with reported results

Propeller diameter in feet

Mean annual wind in MPH

**9-6**  Diameter of a two-blade propeller required to produce an annual average power of 1 kW in the mean annual winds shown.

the *EL* component and reduce the bending moment on the blade. If the rotational speed of the propeller is allowed to follow the wind velocity, this cancellation cannot be complete since the *EL* force is proportional to $V^3$, whereas centrifugal force is proportional to $V^2$. Advantage is taken of this property on modern wind engines such as the NASA Plum Brook unit.

At power, the thrust bearings must withstand the summation of all the *EL* forces and the torque summation of all $MT \times R$ torques. In addition, if the windmill is allowed to slew rapidly in gusts, there is a gyroscopic precession torque in the vertical plane which acts upon the windshaft. The windshaft also must bear the weight of the propeller assembly. In order to restrict the gyroscopic precession forces to safe limits, large modern wind engines are generally not allowed to swing freely with the wind. They are generally moved by a servo-control loop operating from a separate wind direction sensor.

**9-7** The forces acting upon a windmill.

# The windshaft

For the home construction of medium size windmills, a large truck rear axle can be used for the windshaft and pivot bearings. A modern truck axle has the bearings within the hub and is designed to withstand large eccentric loads. It also comes equipped with a large and effective brake that can be employed with the windmill. The hub is also equipped with a large and sturdy bolt circle for attachment of the propeller.

# The Princeton sailwing

One of the developments in modern aeronautical engineering that give promise for an economical construction technique for medium size windmills is the sailwing technique developed for hang gliders and ultra-light aircraft. Figure 9-8 shows the very simple construction of this airfoil section. A pair of ribs and a leading edge fairing are mounted upon a tubular spar. The fabric covering is pulled around the fairing and held at the trailing edge by a wire. The use of the tubular spar is especially handy for fine adjustment in pitch. Of course, the major pitch adjustment is built in with the relationship between the angles of attack of the inner and outer ribs. This feature would also be handy for feathering of the rotor during high winds that would exceed the capacity of the machine.

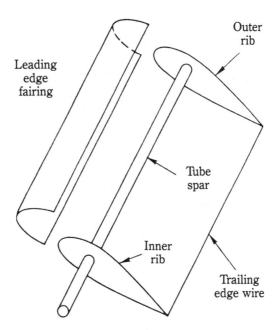

**9-8** The Princeton sailwing construction.

# Variable pitch and feathering

The addition of a variable pitch mechanism is a rather expensive feature to add to a wind engine. It is attractive in that it extends the range of wind speeds over which the machine will operate effectively. However, the entire pitch curve of the helix would have to be variable to be truly effective. In general, this is not done because a mechanism for changing the twist of the blade would be difficult to produce. With an airfoil such as the sailwing, it is a possibility since it would be necessary to rotate only the outer and inner ribs independently. But this would entail some stretching of the trailing edge wire and the fabric. If an automotive type of alternator is used, the maintenance of a constant speed is not required. In this case, variable pitch is not of much advantage.

On the other hand, in the matter of stowage during storms, the ability to feather the propeller would be a distinct advantage. When the propeller is stationary, the wind resistance of the blades is about equivalent to that of a flat plate of the same area and will increase as the square of the wind velocity. If the helical blade can be turned to the most favorable angle, the wind resistance of the blade can be reduced to about a fifth of the resistance in the normal pitch condition. Since the bending moment of the blade at the hub is approximately proportional to the cube of wind velocity and the propeller does not have the advantage of centrifugal force in the stationary condition, blade failure in high winds in the locked rotor condition is a major problem on large wind engines.

# Windmill design

As noted earlier in the chapter, unless you live in an exceptionally windy area, a windmill capable of providing the electrical needs of a modern household is a pretty large device. Furthermore, the high-speed propeller must be pretty sophisticated. At six times a wind speed of 40 mph, the tips of the propeller will be turning 240 mph, and the forces on the tower will be measured in tons. The design of such machines is definitely not the sort of task to be allocated to someone not thoroughly versed in mechanical engineering. A 25-foot diameter propeller disintegrating at high speed from centrifugal force could do an immense amount of damage and endanger life and limb. For this reason, no "cookbook" design or construction details have been offered here. Engineers competent to design such a machine would not need them and people who do would be better off obtaining professional assistance in the matter.

# 10
# Water engines

THE WATER MILL IS—IF ANYTHING—OLDER THAN THE WINDMILL. AN epigram by Antipater of Thessalonica (ca. 85 BC) describes a mill operated by water. Vitruvius describes the Roman mill with a horizontal axle, a gear train, and a millstone set in *De Architectura*, written about 15 BC. The Chinese record the use of a horizontal waterwheel to operate a bellows for a blacksmith's forge in 31 BC. For those interested in a historical and artistic background of the water mill—and the windmill as well—I would recommend *Windmills and Watermills* by John Reynolds, put out by Praeger Publications, NY. From these early beginnings, men have extracted energy from falling water. Unlike the windmill, the use of water power continues on a large scale to the present day, although the rustic mill has been replaced by the hydroelectric power plant.

The vertical axis, horizontal turning waterwheel is usually considered to be older than the horizontal axis vertical turning wheel. The former was usually termed a "Greek" wheel while the horizontal axis machine was termed a "Roman" wheel. Surprisingly, the modern hydroelectric plant is more closely related to the older Greek design in the majority of cases. The reasons for this will be discussed shortly.

## Head and flow

The energy available from water in the normal sense (not including thermonuclear fusion) is determined by the distance the water can be allowed to fall through the machinery, or the *head*, and the amount of water that is available for processing through the machine, called the *flow*. Significant amounts of energy can be extracted from a small amount of water falling a small distance. The machinery differs for the efficient extraction of the energy.

An example is shown in Fig. 10-1, labeled the "Greek turbine." In this unit, the water was allowed to fall down a chute in order to exchange some of its potential energy (due to height) for kinetic energy (due to velocity). At the end of its fall, it impinges upon the blade or blades on one side of a wheel fitted with nearly flat blades. The force or reaction pushed the blade around, and the water drained out below.

**10-1**   The Greek turbine.

There are some obvious losses with this type of unit. First of all, the exchange of potential energy for kinetic energy is not terribly efficient since some of the energy is lost in turbulence and friction in the chute. Secondly, some of the kinetic energy is wasted in splashing off the blades. Thirdly, the blades can never absorb all the kinetic energy since this would bring the water to a full stop and flood the unit. Finally, the unit must give away some head to allow the water someplace to fall.

The Irish turbine shown in Fig. 10-2 can alleviate a few of these problems. It receives the water on a nearly horizontal trajectory. The curved shape of the blades will slow the water to a near stop and eject it horizontally. Since the flow is smoother, it can extract a higher percentage of the kinetic energy. It also gives up less head to clear the water.

A somewhat more modern approach is shown in Fig. 10-3. In this machine, an outboard motor lower unit—complete with gearbox for right angle drive—and a venturi inlet pipe is used. The casing is kept deliberately large to minimize friction losses. Just before the propeller, the tube necks down into a venturi to increase the flow speed before being forced through the propeller. A typical 5-hp lower unit will provide an output on the order of 3-hp at 3600 rpm when operating under an 11-foot head and discharging about 252 lb of water per second.

**10-2** The Irish turbine.

Water

Casing

Water

Shaft

Outboard
motor
lower
unit

**10-3** A machine using an outboard motor lower
unit with a gearbox for right angle drive and
a venturi inlet pipe.

There are several attractive features about this. First of all, the power is on the
right order for a home electrical installation with the 3 hp being on the order of 2.2
kW and secondly, the speed is appropriate for driving a 60-Hz alternator without
gearing. Also, the outboard lower unit is generally available at a very reasonable
price if purchased used on an otherwise defunct outboard motor.

The real class in this department belongs, of course, to the various impulse
wheels designed specifically for hydro power installations.

Figure 10-4 shows the cross section of a Pelton wheel bucket. This unit in three dimensions looks very much like the fully opened shell of a clam. The stream of water from the nozzle is split into two, and the curvature of the buckets is such as to extract the maximum amount of available energy from the water. The clam-shell analogy is complete to the inclusion of the cusp or gap between the shell halves. This is done to minimize the blanketing of the active blade by the next following blade until it is in a position to take over the stream.

Velocity

Water

Nozzle

**10-4**   The Pelton wheel bucket in cross section.

In a well-designed turbine, the efficiency can approach 80 percent compared to the 60 percent figure given for the outboard propeller unit. Actually, a propeller unit can be designed for high efficiency also, but the design and fabrication of such units is a very sophisticated process. The demand for units in the 3- to 10-hp class for home use is so limited that it would be a doubtful investment in engineering time and money.

The principal characteristic of these devices is the fact that nearly all of them require a substantial head in order to operate well. A typical fall for a stream or river is on the order of a foot or two per mile. If we were to impound the waters of such a stream in order to obtain a 12-foot head, it would be necessary to back the waters up for 6 to 12 miles. If the river were not in a deep gorge, this might require the construction of embankments to contain the waters. Obviously, the cost of the land that would consequently be flooded and the dam and embankments would have to be recovered from the revenues or savings on the energy thus acquired. Usually this type of installation is justified only in hilly areas where the pitch of the stream is great and where the stream is contained within a natural gorge. Since this represents something of a special case in geography, it follows that such medium head installations are not very common.

# Energy in falling water

Before going further, it might be instructive to consider the amount of energy that can be obtained from falling water. We shall start by considering what must be about the simplest water engine possible. Figure 10-5 shows a bucket attached to a rope wound around a drum. The bucket contains one cubic foot of water, which—in the case of fresh water—weighs 62.4 pounds. The bottom of the bucket is 10 feet above the ground. The bucket has a potential energy of:

$$62.4 \text{ lb} \times 10 \text{ feet} = 624 \text{ foot-pounds}$$

If friction is neglected, this energy could be delivered through the rotating shaft to an external load.

**10-5** The simplest water engine.

If the external load were to constrain the descent of the bucket so that it descended the 10-foot head in 10 seconds, the output rate would be:

$$\frac{624 \text{ ft-lb}}{10 \text{ sec}} = 62.4 \text{ ft-lb/sec.}$$

This is equivalent to:

$$\frac{62.4 \text{ ft-lb/sec}}{550 \text{ ft-lb/ sec/ horsepower}} = 0.113 \text{ horsepower.}$$

This is in turn equivalent to 84.6 watts. Our more common unit in electrical energy is the kilowatt-hour (kWh), and 1 kWh = $2.656 \times 10^6$ ft-lb.

Next, let us imagine that we are able to use our 10-foot head in some perfect, lossless machine to generate electricity on a continuous basis. Since at the 10-foot head each cubic foot has a potential energy of 624 ft-lb, we would require:

$$\frac{2.656 \times 10^6 \text{ ft-lb/ kWh}}{624 \text{ ft-lb/ ft}^3} = 4256 \text{ ft}^3/ \text{ kWh (at 10-foot head)}$$

With a 1-foot head we would require 42,560 cubic feet, and with a 100-foot head we would require only 425.6 cubic feet.

If the water were coming from a pond and we were to require that the drainage lower the pond by no more than 1 inch, the pond would require an area of:

$$\frac{4256 \text{ ft}^3}{1/12 \text{ ft}} = 51,072 \text{ ft}^2 \text{ (at 10 foot head)}$$

which is:

$$\frac{51,072 \text{ ft}^2}{43,560 \text{ ft}^2 \text{ per acre}} = 1.17 \text{ acres}$$

For the 1-foot head we would require 11.72 acres, and for the 100-foot head we would require 0.117 acres.

Now we would probably like to replenish our water supply by means of rainfall, and the rain does not fall evenly in all places. On a national average basis, the rainfall can be as little as 3 inches per year in Egypt and Arabia and as much as 119 inches in Samoa and the Solomon Islands. However, the rainfall in Rochester, NY, averages 36 inches per year, so let us consider that figure. Because there are 8,766 hours per year the 36-inch rainfall works out to an average of $4.11 \times 10^{-3}$ inches per hour or $3.42 \times 10^{-4}$ feet per hour. To obtain our 4,256 ft$^3$ of water, we would have to drain $1,244 \times 10^7$ ft$^2$ or 285.5 acres or 0.835 square miles. (There are 342 acres per square mile). And all of this would provide only an average of 1 kilowatt provided that we could have a 10-foot head.

The message in this little study is simply that there is more than one house per 0.8 square mile in the area of Rochester, NY, which has an above-average rainfall. We cannot all obtain our electricity from the rainfall. As a matter of fact, the Rochester Gas and Electric Co. does have a hydro plant on the Genesee River in the heart of Rochester. The river is not small and the company is blessed with an 85-foot waterfall. Roughly half of the river goes through their turbines. However, the hydro installation, which is on-line more than 98 percent of the time, manages to produce only about 5 percent of the electricity used in the area.

# Losses

In actual practice, only a fraction of the energy from rainfall is available. First of all, the soil absorbs a healthy fraction, which is largely given up by plants to the atmosphere. The reservoirs also lose a good bit to evaporation. In addition, there are losses of head, which are required for the water to flow to the power station. Harking back to Fig. 10-5, we note that the bucket was filled at the top but that it is the bottom of the bucket that hits the ground first so that of the total head actually available, we lose the height of the bucket. Even if this simple machine had no losses from friction, we could obtain only a part of the available energy.

# The overshot wheel

The overshot wheel is the picture that comes to mind for most people when they hear the term *waterwheel*. However, the traditional overshot wheel was a very inefficient device. As can be seen from Fig. 10-6, the trough filled the buckets at the top. However, as soon as a bucket began to turn from the top position, the water would commence to spill out. By the time the bucket was level with the axle, the water was entirely gone. As shown on the graph to the right, about half of the water was used for about half of the diameter; thus, the wheel could extract only about 25 percent of the energy.

**10-6** The overshot wheel.

This loss can be partly eliminated by "feathering" the wheel as shown in Fig. 10-7. The buckets on a feathered wheel are held upright until they near the very bottom of the wheel, whereupon a mechanism dumps them. A wheel of this sort must turn very slowly in order to avoid losing the water as a result of centrifugal force. It must therefore be rather large.

Suppose, for example, that we wish to construct a wheel to give us 2.2 kW (3 hp) with a 10-foot effective diameter. We would need to handle 4,256 ft³/hr × 2.2 = 9363 ft³/hr. If the wheel turned at 10 rpm or 600 revolutions per hour and the wheel had 35 buckets, the water would be handled at the rate of 600 × 35 = 21,000 buckets per hour and each bucket would have to have a capacity of 0.446 cu ft. If

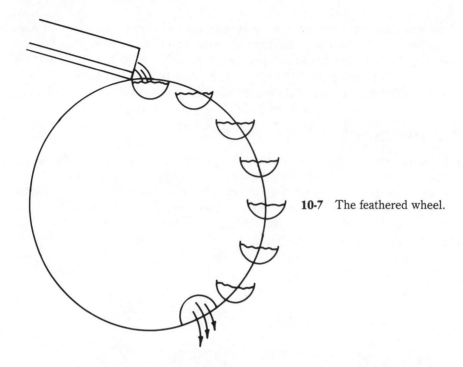

**10-7**   The feathered wheel.

the buckets were semicircular and 1 foot in diameter, they would have to be about 14 inches wide. The overall diameter of the feathered wheel would have to be something like 11 feet to get a 10-foot effective drop and still clear the water on the bottom. This is a pretty substantial sized piece of hardware compared to the propeller arrangement with the same rating of Fig. 10-3.

When the trough clearance for filling and the bottom clearance are added, a head of about 12 feet would be required in order to operate the unit. In terms of overall efficiency, the unit would use less water to produce the same power that the propeller did. However, it would require gearing up by a factor like 360 to attain sufficient speed for an alternator.

# Low-head water engines

In Europe and England, the use of the undershot wheel was far more common than the use of the overshot wheel for the simple reason that there are far more sites suitable for a low-head installation than there are for a medium-head installation. It is far easier to obtain a 1-foot head than it is to obtain a 10-foot head. A number of these were located at the mouths of tidal estuaries and ran in one direction when the tide was coming in and the other when it was going out.

Figure 10-8 shows the working portion of an undershot wheel. The wheel is fitted into a close-fitting millrace to minimize losses around the sides and bottom. The 1 paddle (or float, as they are frequently called on an undershot wheel) has a definite difference in head between the 2 sides. At the surface, the pressure is zero

**10-8**   The undershot wheel.

and climbs linearly at the rate of 62.4 lb/ft$^2$ per foot of depth. The average pressure is 1/2 of the maximum pressure so that if the head difference is 1 foot the average pressure is 31.2 lb/ft$^2$.

Now let us assume that we would like to build a 3-hp undershot mill and that the effective radius of the wheel is 9.5 feet and the wheel is turning at 10 rpm. At the effective radius point, the wheel speed works out to be 9.95 feet per second. The work on the wheel is:

31.2 lb/ft width $\times$ 9.95 ft/sec = 310 ft-lb/sec/ft width. For 3 hp = 1650 ft-lb/sec, we obtain a width of 5.32 ft. The mill will be passing 9.95 ft/sec $\times$ 5.32 ft $\times$ 1 ft $\times$ 3600 sec/hr = 190,562 ft$^3$/hr. Compared to the theoretical value for a 1-foot head of 43,560 ft$^3$/hr/kWh this works out to 50-percent efficiency.

In an actual installation, the efficiency would not be this good because leakage around the floats would reduce the effective head and the water passed through would not immediately depart. Therefore, it would cause a certain amount of back pressure on the wheel. A practical 3-horsepower wheel would probably be about 8 feet wide and 20 feet in diameter. This is an awfully large piece of equipment compared to the outboard unit. However, it will recover "free" energy in a very low density situation. In actual practice, it would probably be necessary to have something like 1¹/₄ or 1¹/₂ feet of head to allow for the millrace entrance and exit. A wheel such as this will actually operate in an open stream without the fitted millrace, but the power is greatly reduced.

# The breast-shot wheel

For installations where the head is more than a foot but not as much as 10 feet, the breast-shot wheel provides an optimum solution. As may be seen from Fig. 10-9, the breast-shot wheel combines some of the features of the overshot wheel and the undershot wheel. Because of spillage, it gets only about 25 percent efficiency. However, it will operate at medium heads quite effectively. With end plates (not shown in the drawing), the breast-shot wheel does not require a close-fitting millrace for effective operation. For many years, the breast-shot wheel was the wheel of choice for low-head installations.

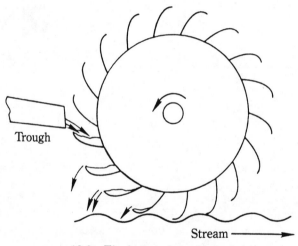

Trough

Stream ⟶

**10-9**   The breast-shot wheel.

# Other low-head types

From the 1880s well into this century, there were a number of manufacturers producing special turbines that had marvelously scrolled and fluted cases and runners intended for producing water power from low-head installations. Unfortunately these units seem to have disappeared completely from the marketplace because they became impractical for industrial applications. Most of these designs are far too complex for most home workshoppers. One of these units might be found in an old factory in operable or repairable condition provided the owner does not want it for a museum piece.

# High-head installations

In hilly or mountainous locations, one often observes heads of 100 or more feet. As noted earlier, the higher the head, the smaller the flow required. The pressure of water increases by 62.4 lb per ft$^2$ per foot head or 0.433 lb per in$^2$ per foot head. At a 100-foot head the pressure is 43.3 lb per in$^2$, which is well within the rating of the oil pumps used on diesel trucks. These pumps are also generally of the positive displacement type and will run equally as a motor. Ratings in the 1- to 3-horse-power range are not unusual for large diesel engines so that these units may generally be adapted to home electrification. They are not generally constructed of corrosion resistant alloys so that pump life might be limited to a few years. If the price is reasonable, though, this might not be a severe drawback. A commercially made water pump of suitable pressure and power rating may also be selected from a number of vendors.

# System considerations

A basic hydroelectric system consists of the following items:

- A source of water
- A place to dispose of the water
- A strainer or grating to keep debris out of the system
- A valve or gate to permit system shutdown
- A water engine of some sort
- A generating system to produce the electricity
- An electrical control system for control of voltage and frequency
- An overflow or bypass mechanism so that heavy rains do not cause flooding

Starting with the matter of the valve or gate for system shutdown, it is usually advisable to place this at the "high" or "entry" point of the system. When placed at the top, this valve may be shut and the system drained for maintenance. This is particularly required in high-head systems, but even on an undershot wheel, it is nice to be able to dry the millrace for maintenance work on the wheel.

In the case of the generating system a choice sometimes exists. When it is easy to regulate the speed of the wheel by means of the inlet gate, it is often possible to directly drive a 60-Hz alternator and accurately control the speed. This is probably the least expensive and most direct approach. In cases where the wheel speed is likely to be widely varying, it is probably easiest to directly drive a dc generator and then use a motor-generator set to obtain the requisite 60 Hz for the final usage. This is expensive because it requires three electrical machines, rather than one. However, it is probably more effective than the use of a variable speed mechanical transmission. If the entire system head is not used at full load, it is usually possible to produce a servo-control for the inlet gate to hold the output speed constant under varying load. This is the approach used in large commercial hydro installations.

The overflow or bypass system will often consist of a weir cut into the top of the impoundment dam with a bypass channel dug to permit excess water to flow around the powerhouse and directly into the tailing stream. In areas where heavy spring rains are the rule, care must be taken in the design of the bypass to ensure that it is adequate to handle the heavy runoff. Also the tailing stream channel must be large enough to handle the runoff so that is doesn't rise so high that it backfills the engine runoff. This backfilling can seriously reduce the output of the machine since the water must be able to fall freely out of the machine to provide a reasonable flow. Of course, an even worse case exists when the tailing stream rises high enough to flood the powerhouse. This can lead to very expensive damage to the electrical components, and the flooding will usually leave the powerhouse filled with mud. Worse still is the possibility that a really severe flood will wash away the whole powerhouse.

The greatest danger of all is that a series of heavy rains can weaken and perhaps rupture the impoundment or the dam. In this case it is likely that the entire contents of the impoundment will be released in a single tidal wave. The destructive force of a 10-foot wall of water rushing down a gorge or stream is almost beyond belief. Bridges, culverts, houses, or nearly anything that gets in the way can be swept aside like matchsticks.

*A water impoundment of any significant size should never be erected without the services of a competent civil engineer. Local laws, ordinances, and design codes should be rigorously observed and investigated before any such construction is initiated.*

# A home hydroelectric installation

Figure 10-10 shows some of the principal features of a small medium-head home hydroelectric installation. The unit uses a propeller motor of the sort shown in Fig. 10-3. At the water inlet, a trash rack is placed to keep branches, rocks, and large debris from falling through the motor and damaging the propeller or gears. A belt drive is used to elevate the alternator some distance above the tailing stream to reduce the likelihood of flood damage.

The speed of a propellerlike water engine is a function of the torque load and the head. Therefore, a certain amount of head is given away for purposes of control. A frequency sensor measures the frequency of the load current and operates the gate control motor to adjust the head accordingly. If a heavy load comes on line, this will tend to slow the alternator, thereby producing a phase lag or a reduction in frequency. The control responds by opening the control gate, which raises the head in the casing; this in turn offsets the lag. Voltage and frequency controls can be as described in the section on motor generators. Depending upon the speed of response of the water level in the casing and the inertia of the motor-generator system, it might or might not be possible to operate the system in a true phase-locked condition.

The compressed air system is used to provide a continuous stream of bubbles for an area directly in front of the inlet. This will tend to circulate warmer water from the bottom of the impoundment to the surface and will aid in keeping the inlet ice-free in cold climates. The depth of the impoundment should somewhat exceed the greatest depth of ice generally encountered in the area.

**10-10**  A small hydroelectric installation.

It is probably advisable to install the switchboard and control system in a separate room that is well above the water outlet. That area will generally be completely saturated in humidity, and running condensation on the walls and motor casing is very likely. The presence of so much moisture is undesirable around the electrical equipment. Extreme moisture also raises the likelihood of annoying electrical shocks when the equipment is touched, because it would be nearly impossible to keep the insulators free of moisture.

An installation of this sort, while obviously expensive, will provide about the finest type of home electrical service. It has the advantage of being nearly as dependable as the water supply itself and requires no storage of electrical energy. The system can be easily built, so that very large overloads can be handled temporarily by simply opening the control gate wider during the overload. The requirement for an elaborate prioritizing and load-scheduling system is either eliminated or minimized. In short, if one has the money, geography, and legal feasibility to permit the construction of such an installation, it will represent a premium quality home electric plant.

# Environmental impact

The establishment of a hydroelectric plant always has some significant impact upon the surroundings. In particular, the construction of the impoundment will probably flood a great deal of land. If you do not own this land, it will be necessary to buy it in most cases. It is not unlikely that roads will have to be rerouted and so on. In populated areas this is liable to require a certain amount of political activity.

The people living downstream from the installation are also apt to have a few words to say about the subject. During the initial fill period, the flow of their stream will be either completely shut off or greatly diminished. Once the plant is constructed and filled, they might feel that the dam represents a substantial threat to their life and property because of the tremendous damage that a failure could produce.

In terms of ecological impact, there might also be certain repercussions. During the initial fill period, the reduced flow is very likely to damage or destroy the downstream aquatic life. Someone is bound to find that the stream carries an obscure subspecies like the snail-darter, which could be extinguished by the construction.

Even the lowly undershot wheel, if operated in a tight millrace, represents an impenetrable obstacle in at least one direction for aquatic life. If the stream is used by migratory fish, it might be necessary to provide a fish ladder or similar installation to permit passage of fish around the installation.

By the way, some of the finest fishing anywhere in the United States is found in the huge impoundments built by the Tennessee Valley Authority. These impoundments provide a huge reservation for game fish where no fish previously existed because there were only shallow rivers. In fact, these impoundments abound with bass and other fish, whose ascendency came at the expense of certain riverine types. Whether these changes represent an ecological improvement is probably a matter of viewpoint. It is certainly a subject of debate.

In the long haul, the installation of a hydroelectric plant can provide a nice environment for aquatic life if it is carefully handled. The net flow of the stream will be unaltered. The power is produced with no pollutants and no wastes to dispose. However, the acts of construction cannot but help, at least temporarily, to alter the environment. Because of the sheer size of the project, this might bring vigorous objections from some quarters. There is probably no single answer to the question.

# 11

# Electricity-from-sunshine principles

EVERY READER KNOWS THAT THE EARTH REVOLVES ABOUT THE SUN AT the center of the solar system. However, for the purposes of our discussion here, it is necessary to dig a little bit deeper into our relationship with the sun. As will be noted shortly, the details of this relationship have a significant bearing upon the design of any solar-powered electrical system.

The earth revolves about the sun in a nearly round elliptical orbit with the sun at one focus at a mean distance of about 93 million miles. The equator of the earth is inclined to the plane of the ecliptic by an angle of about 23.5°. This relationship can be visualized from the diagram in Fig. 11-1. The direction of the rotation of the earth with respect to the revolution about the sun is such that one revolution is subtracted in each circuit. Stated another way, every 365 days, 6 hours, 9 minutes, and 9.54 seconds of mean solar time, any fixed star will have passed the meridian 366 times. This is the reason for the changing of the constellations with the seasons.

The 23.5° inclination of the equator is even more significant in terms of solar energy and climate. Surprisingly, the foci of the ellipse are located so that the earth is actually slightly closer to the sun on the 21st of December than it is on the 21st of June. However, the slight increase in the received radiation is completely overshadowed by the fact that on December 21 the sun is very low in the sky and we experience the shortest day of the year. The lower part of Fig. 11-1 shows the geometry of this situation. Not only is the day shorter at the high latitudes in the northern hemisphere but the zenith angle attains its highest noontime value. At 44 degrees 3 minutes north on the 21st of December the zenith angle of the sun at noon is 67.60 degrees. The amount of sunshine falling upon a horizontal surface varies as the cosine of the zenith angle, which is 0.381 for 67.6 degrees. The sun rises at about 7:45 AM EST and sets at something like 4:35 PM EST and it is winter in Rochester, NY. At its highest point the sun is only 22.4 degrees above the horizon.

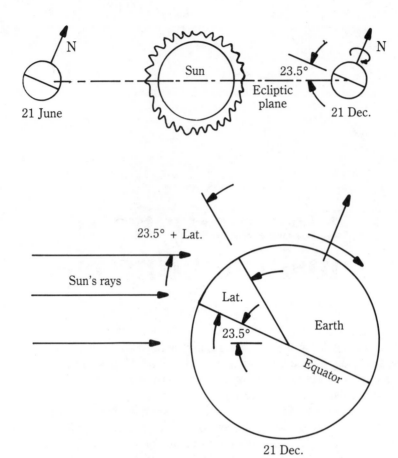

**11-1** Geometry of the sun-earth system.

In contrast, on June 21 the zenith angle drops to 20.75 degrees (90–69.25), for which the cosine is 0.935. On horizontal surface in Rochester there is more than three times as much energy on the first day of summer as there is on the first day of winter. In addition to the cosine law, which governs the incident energy, there is atmospheric attenuation, which is proportional to the sine of the zenith angle. The sine of 67.6 degrees is 0.925 and the sine of 20.75 degrees is 0.354. In December we have 2.61 times as much atmospheric absorption or attenuation as we do in June. You do not need a fancy instrument to tell you that the December sun holds less energy than the June sun. In addition to this the sun rises at 4:30 AM EST and sets at 9:45 PM EST. (We are actually on daylight time so both times would be 1 hour later EDT.)

The matter of latitude makes a very significant difference in the viability of solar power generation. At Bodo, Norway, on December 21 the sun never rises and on June 21 it never sets; this is the definition of the Arctic Circle. One could proba- bly not capture a single hour of useful energy during the months of December and

January. Conversely, at Lagos, Nigeria, a solar electrical system could probably provide useful energy for 11 hours a day all year long (except during storms).

As it affects energy collection, the 23.5 degree tilt of the earth's axis is very significant. If we have some simple solar array, for example, it is possible to split the difference and tilt it so that it is pointed too low in the summer and too high in the winter. The cosine of 23.5 degrees is 0.917 so we would give away 8.3% at the extremes and be just right on September 21 and March 21, the Equinoxes when the sun crosses the equator. However, we shall show shortly that if we choose to operate a heat engine from the sun, it is necessary to use a concentrator to make things hot enough. The concentrator must be accurately aimed. We also see that it must be able to swing through an angle of 47 degrees in elevation and nearly 247 degrees in azimuth.

# Finding the sun

At first glance it seems a bit silly to question where one finds the sun. You can simply look for it. We shall shortly show mechanisms for tracking the sun with a high degree of accuracy. However, these techniques work only when the sun is visible. How about when it goes behind a cloud? Large solar collectors cannot be rapidly slewed because of the stresses imposed by rapidly accelerating a large structure weighing many thousands of pounds. Ideally we would like to be able to keep the collector tracking even when the sun is hidden so that energy can be collected when the cloud passes.

Offhand one might think that it would be simple to arrange some sort of gadget like a clock whose hands run on an axis parallel to the pole. If the hand makes one revolution per day, it should point in the direction of the sun. Well, this would be a start; however, it would not account for the seasonal height changes. In addition to this, there is the little matter of our orbit. If you remember, I noted that the earth's orbit is a *nearly* circular ellipse with the sun at one focus. Well, Kepler's law states that the area swept out by the radius vector connecting the planet to the sun is constant with unit time. This means that when the earth is closest to the sun and the radius vector is shortest, we move the fastest, and when the earth is farthest from the sun and the radius vector is longest we move the slowest. In other words, we move at variable pace over our elliptical path about the sun. This is reflected in the fact that solar noon, when the sun is due south of your location, can get as much as 16 minutes ahead of local mean noon in November and can get as much as 14 minutes behind local mean noon in February. Since the earth rotates 15 degrees per hour this means that the error can be as great as 4 degrees. The error is known as the EQUATION OF TIME.

The distance that the sun is above or below the plane of the earth's equator is called the APPARENT DECLINATION OF THE SUN. Tables of these values are to be found in the Keuffel and Esser *Solar Ephemeris*, which is published annually. A second source for this data is the *US Naval Almanac*. published by the US Government printing office. The K&E book is less expensive and can usually be obtained from drafting supply houses.

# Calculating altitude and azimuth of the sun

The calculation of the altitude and azimuth of the sun involves the solution of some equations in spherical trigonometry. The traditional way to deal with this problem involves an imaginary CELESTIAL SPHERE, which in this case has a radius equal to the earth-sun mean distance. On this scale, it is valid to assume that the earth radius is negligible and that the earth-sun distance is constant. The illustration of Fig. 11-2 shows the geometry of the situation. The size of the earth has been exaggerated greatly so that it can be seen. Imagine that you are standing in your back yard at your potential solar site. You have a surveyor's transit duly leveled and a plumb bob. You are in North America and it is a winter day.

The direction the plumb bob points is toward the center of the earth and beyond that to the NADIR. If the up direction of the plumb bob string is extended to the sphere it reaches it at the ZENITH, which is above the plane of the equator by the amount of your north latitude (south latitudes are considered minus).

One side of the solar triangle is the sun's polar distance. This is equal to 90 degrees plus the declination taken from the ephemeris for the day. Now the declination is continually changing so the ephemeris table quotes it for 0 hours at the zero meridian, which happens to run through the center of the telescope at the observatory in Greenwich, England. The ephemeris then gives a straight-line difference on a per hour basis for correction of the reading for elapsed time since 0 hours on the day in question. If you multiply the correction by 24 and add or subtract it from the day's entry you obtain the following day's entry. It is important to note that the correction as given in the ephemeris is not algebraically signed. Look at the entry for the next day and, if it is more southerly (that is more negative), make the correction negative as well.

The second side of the triangle is the COLATITUDE (PZ). This term is simply equal to 90 degrees minus your north latitude. In Program 11-1 this is given in radians because the Apple does not understand degrees. The polar distance (PS) is computed on line 1510 and the colatitude (PZ) is computed on line 1500.

The third side of the triangle, the ZENITH-SUN DISTANCE, is an item we need to know since the elevation of the sun as seen from our location (HT) is simply 90 degrees minus the zenith-sun distance (AA)

A third element is required in order to solve the triangle, namely, the polar angle of the sun (T). This is the angle between the great circles going through the pole and the sun and the pole and our meridian. The sun moves through the celestial sphere at the rate of 15 degrees per hour so our local time will be behind GREENWICH MEAN TIME (usually quoted as GMT) or UNIVERSAL TIME (UT) by our WEST LONGITUDE in degrees divided by 15 degrees/hour.

Since the 1880s in the United States, we have operated on STANDARD TIME, which is computed at the 75th, the 90th, and the 105th meridians. These are 5, 6, and 7 hours behind GMT. In Rochester, NY, at 77 degrees 31 minutes west, we keep EASTERN STANDARD TIME, which is 5 hours behind GMT. However, with respect to the mean sun, noon occurs in Rochester 10 minutes 4.2 seconds later than in Philidelphia on the 75th meridian and 49 minutes 55.8 seconds before St.

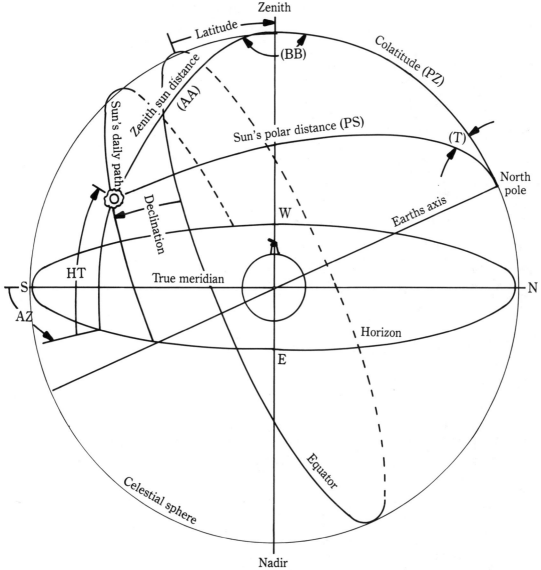

**11-2** The celestial sphere.

Louis on the 90th meridian. In the summer EASTERN DAYLIGHT TIME is 6 hours behind GMT.

The program calculates the difference between mean local time and GMT in line 1310, and the polar angle (T) is computed on line 20. In this computation the corrected EQUATION OF TIME value (HA) is added to transform the time value to local sun time from local civil time.

### Program 11-1. Altitude and azimuth of the sun, by Jack Kuecken

```
100   REM ***********************************************
110   REM            ALTITUDE AND AZIMUTH OF
120   REM                  THE SUN
130   REM            BY JACK KUECKEN
150   REM ***********************************************
160   HOME
165 PI = 3.14159265
170   INPUT "ENTER DATE ";D$
180   GOTO 1000
190   REM  *********************************************
200   REM       THE DEGREE/MIN/SEC ROUTINE
210   INPUT "DEGREES ";A1
220   INPUT "MINUTES ";A2
230   INPUT "SECONDS ";A3
240   IF A1 < 0 THEN A2 =  - A2:A3 =  - A3
250 A = A1 + (A2 / 60) + (A3 / 3600)
260   RETURN
390   REM *********************************************
395   REM        CORRECTION TO HOURS
400   INPUT "HOURS ";H1
410   INPUT "MINUTES ";H2
420   INPUT "SECONDS ";H3
430   IF H2 < 0 THEN H3 =  - H3
440 H = H1 + (H2 / 60) + (H3 / 3600)
450   RETURN
490   REM *********************************************
495   REM           THE CAPTION ROUTINE
500   PRINT D$
510   PRINT "AT ";LA;" DEGREES NORTH AND ";LO;" DEGREES WEST"
520   PRINT " THE SUN AZIMUTH AND ELEVATION ARE:"
530   PRINT ""
540   RETURN
990   REM ***********************************************
995   REM       THE MAIN ROUTINE
1000   PRINT "ENTER NORTH LATITUDE ": GOSUB 200
1010 LA = A
1015   PRINT LA
1020   PRINT "ENTER WEST LONGITUDE ": GOSUB 200
1030 LO = A
1035   PRINT LO
1040   PRINT "NOTE:THE DIFFERENCES IN DECLINATION"
1045   PRINT "AND THE EQUATION OF TIME ARE NOT "
1047   PRINT "ALGEBRAICALLY SIGNED IN THE TABLES"
1049   PRINT ""
1060   PRINT "ENTER APPARENT DECLINATION OF ": PRINT "THE SUN AT
       0 HRS GMT": GOSUB 200
1070 D = A
1075   PRINT D
1080   PRINT "ENTER DIFF. IN DECLINATION/HR"
1090   INPUT DD:DD = DD / 60
1100   PRINT "ENTER EQUATION OF TIME "
1110   GOSUB 410
1120 HA = H
1125   PRINT HA;" HOURS"
```

```
1130   PRINT "ENTER EQN DIFFERENCE FOR ONE HOUR"
1132   PRINT "OBSERVE SIGN ON THIS ENTRY"
1140   INPUT DT:DT = DT / 3600
1150   PRINT "ENTER GMT (24 HR FORMAT)"
1160   GOSUB 400
1170 GT = H
1175   PR# 1
1176   GOSUB 500
1180   PRINT "FOR GREENWICH TIME= ";GT
1190 HA = HA + (DT * GT)
1200 D = D + (DD * GT)
1210   REM   HA IS CORRECTED EQ'N OF TIME
1220   REM   D=CORRECTED DECLINATION
1290   REM *******************************************
1300   REM      CORRECT TO LOCAL TIME
1310 CT = GT - (LO / 15)
1320 T = (CT - 12 - HA) * 15
1350   REM   THIS IS THE POLAR ANGLE
1360   REM   BETWEEN YOUR MERIDIAN AND THE SUN
1490   REM *******************************************
1495   REM         NOW DO THE TRIANGLES
1500 PZ = (PI / 2) - (LA * PI / 180): REM   THE ZENITH POLAR DIS
    TANCE
1510 PS = (PI / 2) - (D * PI / 180): REM   THE SUN/POLE SIDE
1520   REM   NOW SOLVE THE TRIANGLE FOR THE THIRD SIDE
1530 CA = ( COS (PZ) *  COS (PS)) + ( SIN (PZ) *  SIN (PS) *  COS
    (T * PI / 180))
1550 AA =  - ATN (CA /  SQR ( - CA * CA + 1)) + PI / 2
1555   REM   THIS IS THE ARCCOS(CA) FUNCTION
1560 HT = (PI / 2) - AA
1570   PRINT "THE ALTITUDE OF THE SUN IS ";(HT * 180 / PI);" DEG
    REES"
1690   REM *******************************************
1695   REM         NEXT FIND AZIMUTH
1700 CB = ( COS (PS) - ( COS (PZ) *  COS (AA))) / ( SIN (PZ) *
    SIN (AA))
1710 BB =  - ATN (CB /  SQR ( - CB * CB + 1)) + PI / 2
1720 AZ = PI - BB
1730   IF CT < 12 THEN AZ =  - AZ
1740   PRINT "AZIMUTH IS "(AZ * 180 / PI);" DEGREES FROM MERIDIA
    N"
1750   PRINT ""
1760 GT = GT + .5
1770   GOTO 1180
10000   STOP
```

# Calculating the altitude

With the angle ($T$) known, we have two sides and an included angle so we may compute side ($AA$) using the formula:

$$\cos(AA) - \cos(PZ)\cos(PS) + \sin(PZ)\sin(PS)\cos(T)$$

This computation is performed on line 1530. The business on line 1550 is a constructed function to do the arccos manipulation, which is not included in the Apple although it is to be found on most engineering calculators. The height of the sun at your location is simply 90 degrees minus (AA) as performed on line 1560, in which (HT) is computed.

# Finding the azimuth

We now know three sides and an angle of the spherical triangle so we have the necessary information to compute angle (BB). It should be relatively obvious that (BB) is the complement to the angle between the meridian and sun (AZ) on the figure. This computation is performed on line 1700, which gives us the cos (BB), and angle (BB) is computed with the arccos routine on line 1710.

The program is arranged to increment the time and print another entry on line 1760.

# The program output

The data of Table 11-1 show typical results computed for the end of January when the sun is low and the days are short in Rochester. Note that the sun is below the horizon at 12:00 GMT. Local noon occurs at 17:06:50 GMT, which is 12:06:50 EST, and the maximum height of the sun is about 30.4 degrees. At sunrise the sun is 66.7 degrees east of the meridian, as shown by the minus reading.

A purist would probably apply a correction for refraction to the readings; however, above 10 degrees this is less than 5 seconds of arc, which is probably negligible for most solar tracking purposes. It is seldom practical to operate a solar thermal plant at elevations less than 10 degrees. The radiation is generally too feeble for practical operation.

### Table 11-1.

---

**January 31, 1984**
**At 44.05 degrees north and 77.5166667 degrees west**
**the sun azimuth and elevation are:**

For Greenwich time = 11
The altitude of the sun is –11.5124092 degrees.
Azimuth is –76.6528979 degrees from meridian.

For Greenwich time = 11.5
The altitude of the sun is –6.22296132 degrees.
Azimuth is –71.7481033 degrees from meridian.

For Greenwich time = 12
The altitude of the sun is –1.07246651 degrees.
Azimuth is –66.7355729 degrees from meridian.

For Greenwich time = 12.5
The altitude of the sun is 3.89621941 degrees.

**Table 11-1. Continued.**

Azimuth is –61.5421755 degrees from meridian.

For Greenwich time = 13
The altitude of the sun is 8.63493188 degrees.
Azimuth is –56.0983304 degrees from meridian.

For Greenwich time = 13.5
The altitude of the sun is 13.0887468 degrees.
Azimuth is –50.3383057 degrees from meridian.

For Greenwich time = 14
The altitude of the sun is 17.1949778 degrees.
Azimuth is –44.2027555 degrees from meridian.

For Greenwich time = 14.5
The altitude of the sun is 20.8830068 degrees.
Azimuth is –37.6439677 degrees from meridian.

For Greenwich time = 15
The altitude of the sun is 24.075532 degrees.
Azimuth is –30.6340328 degrees from meridian.

For Greenwich time = 15.5
The altitude of the sun is 26.6919305 degrees.
Azimuth is –23.1752492 degrees from meridian.

For Greenwich time = 16
The altitude of the sun is 28.6542373 degrees.
Azimuth is –15.310404 degrees from meridian.

For Greenwich time = 16.5
The altitude of the sun is 29.8955427 degrees.
Azimuth is –7.12861028 degrees from meridian.

For Greenwich time = 17
The altitude of the sun is 30.3693861 degrees.
Azimuth is –1.23829825 degrees from meridian.

For Greenwich time = 17.5
The altitude of the sun is 30.0575452 degrees.
Azimuth is 9.63147142 degrees from meridian.

For Greenwich time = 18
The altitude of the sun is 28.973447 degrees.

---

# Just how much energy does the sun give us?

The effective temperature of the sun as measured from the earth is 6150°K. The effective subtended angle of the diameter of the sun as viewed from the earth is a little more than a half-degree and the diameter of the earth viewed from the sun is $4.09 \times 10^{-3}$ degrees. The earth intercepts about one two-billionth of the energy radiated by the sun. Of this fraction about half is radiated back into space by our atmosphere. We actually receive about $10^{18}$ kWh per annum. About 30 percent of this is reflected into space by plants during the daytime. The rest is distributed about as follows:

| | |
|---|---|
| Nighttime radiation | $.7 \times 10^{17}$ kWh |
| Evaporation of water | $4.5 \times 10^{17}$ kWh |

On the average, plants utilize only about a thousandth of the solar energy incident upon them. Land vegetation consumes about $4.1 \times 10^{13}$ kWh and marine vegetation consumes about $33 \times 10^{13}$ kWh for a total of $3.7 \times 10^{14}$ kWh. On the average, one kWh of solar energy can produce about 0.067 lb of vegetation and the same amount of oxygen. It is estimated that the largest amount of solar energy that could be produced from a site in Arizona is 328,000 kWh per acre per annum. If the same area were farmed and the vegetation burned to produce electricity, it is estimated that the yield would be about 44,700 kWh per acre per year. In terms of electrical energy production the direct solar process is more than 7 times as efficient. The burning of the plants would just about consume the oxygen produced by the photosynthesis of the plants; the burning would just about liberate the same amount of $CO_2$ as had been absorbed.

The 328,000 kWh per acre per annum is an interesting figure since this works out to an average output of 37.42 kW. At 1.6 kW per home, this would support 23 homes. Even if the output in Rochester were only 10 percent as great, this would be enough electricity to support more than two homes per acre, which is about the density of the suburban areas. It has been estimated that if all the wind in the world that was practical to employ were put in use, it would supply only about 3 percent of our current electrical demands. Similarly if all the water in the world practical for hydro power generation were employed, it might meet about half of the world energy demand. In the US it is estimated that about 40 percent of all the practical hydroelectric power is currently running, and it produces about 5 percent of our electricity. If all the practical hydropower in the US were put into service, the share for hydro would rise to only 12.5 percent.

The rest of the world is not better blessed in terms of potential hydropower; it is simply that the rest of the world uses so much less electricity per capita. The US is considered to have 13 percent of all the potential hydropower in the world and of that fraction some 40 percent is up and running right now. As the standard of living rises through the remainder of the world, the shortfall of hydropower compared to world demand will probably increase.

Among the renewable resources, only in solar power do we find the potential for an energy source capable of supplying more energy than is used.

If we define the standard of living of people as being related to their energy consumption, then the US currently has about a 2.4 kW per capita standard of living. This is just about split equally three ways between electricity, transportation, and heating (for homes, industry, and all other purposes). Suppose that of the $4.5 \times 10^{17}$ kWh per annum that is used by the earth to evaporate water from the oceans we were to acquire just 0.1 percent or $4.5 \times 10^{14}$ kWh per annum. Dividing by the hours in a year gives a continuous yield of $2.90 \times 10^{10}$ kW. This would supply 2.4 kW to 12.1 billion people, which is more than three times the estimated world population. With a relatively small reduction of energy input into the oceans, we could more than provide a 2.4 kW standard of living for every man, woman, and child on the face of the earth!

However, before we become too ecstatic about this possibility, let us consider some of the ramifications. The area of the earth is about $6.6 \times 10^7$ square miles, and of this some $5.3 \times 10^7$ square miles is ocean. A tenth of one percent of the oceans is $5.3 \times 10^4$ square miles. If we were to assume that the system had an overall efficiency of 33 percent and that we had three floating power islands in the Atlantic, Pacific, and Indian Oceans to provide round-the-clock power generation, then each island would have to be 230 statute miles on a side. This is larger than a great many of the countries on the earth! Just figuring a way to assemble, regulate, and distribute the power from these behemoths would require the work of generations of engineers. The construction would require a significant fraction of all the iron and copper on earth, and there probably isn't that much wealth. In addition to this, the ecological impact of shadowing the waters in an area 230 miles on a side is at the very least a subject for speculation. At the very least, the local climate could be expected to vary, and there are some who maintain that violent climatic changes like an ice age can be triggered by effects no larger.

Perhaps even worse is the political aspect of such a concentration. With the entire world dependent upon the supply of electricity of such an organization, the opportunity for tyranny would be unprecedented. Any nation could be instantly brought to a grinding halt by simply turning off its electricity—its sole source of energy. Within a matter of days, the lack of transportation, refrigeration, water pumping, heating, and lighting would bring starvation and chaos. The presence of such a disciplinary weapon would probably bring instant acquiescence at even the mention of a threat. Any attempt to construct such a worldwide power generation and distribution system would probably be met with a huge and violent opposition on the political front.

From the standpoint of the utility companies, the use of solar energy has a questionable practicality. In the perpetually sunny southern locations, the energy is available during only a portion of the day. In the northern climates, it might be essentially absent for weeks or months at a time. The possibility exists for electrical energy to be shipped over large distances. However, very large power transmission lines are expensive and difficult to operate. They also tend to be lossy. Current efforts to construct large intertie lines have been plagued by political opposition from environmental groups.

As a practical matter, electricity in commercial quantities is completely volatile. It must be consumed as it is made. With the exception of pumped water storage, there is currently no practical way of storing electrical energy. At the present writing, the facilities for pumped storage tend to run an overall efficiency of about 60 percent and have an installed kVA rating of less than 1 percent of the national average requirement.

Because the entire hemisphere is dark at times, this boils down to the requirement of duplicating the complete generating facility. Since solar plants are estimated to run $12,000 (installed) per kVA versus $12 to $120 for fossil or nuclear plants, this would mean a large multiplication of the capital investment to provide a facility that would be useful mainly for assuming the peak loads caused by air conditioners. Unless or until an economic mechanism for bulk power storage is perfected, the generation of large quantities of solar electrical power will be confined to very sunny locations.

A similar analysis applies for home solar electricity, particularly in the northern climates. Solar power is present during only a fraction of the day in any location. In the northern climates, the fraction might fall to a quarter of the hours and the energy available might fall to something less than 10 percent on clear days and to zero on cloudy days. In Rochester, NY, it seems likely that the months of November through April would yield only about 3 to 4 percent of the electricity available from a given installation during the months of June through August. For such areas, it seems likely that a storage system capable of capturing energy in June and holding it until next April or May would be a requirement for a solar-only system. Of course, areas like Arizona, New Mexico, and Utah would require storage for only a matter of a half-day or day. Places like California and Florida, which have bright sunshine year round and a southerly location but are afflicted with rainy periods of a week or so, would require something on the order of a week of storage for a solar-only system.

# Solar energy recovery

As noted toward the close of chapter 4, the present monocrystalline solar cells are too prohibitively expensive to permit their use as a practical source of solar energy in the home. If some form of focusing or concentrating mechanism were applied, the yield per cell could be improved to some extent. However, the poor efficiency of the cells means that the majority of the energy incident upon the cell is wasted in heat. Because the cells themselves can easily be destroyed by overheating, the amount of improvement attainable in this direction is very limited. In weak sunshine or on a cloudy day, the focusing mechanism can be used to bring the output of the solar cell back to about the level that it would have had in bright sunshine. Yet in bright sunshine, any significant increase in output due to focusing would be bought at the cost of overheating and damage to the cell.

If the solar cell is too expensive, then how may we turn the energy contained in sunshine into electricity? Although a number of experiments are being pursued in which photochemical reactions may be used to "fix" solar energy into some form of fuel, at the present writing there are only two established and practical mechanisms for transforming solar energy into electrical energy:

- Using the sunshine to grow plants, which may subsequently be used as fuel in a heat engine.
- Using the solar energy directly to operate a heat engine.

# The heat engine

In order to understand some of the rationale behind the discussions to follow, it is necessary that we have some understanding of the properties of heat engines in general, since these have a great deal to do with the construction of a usable solar plant. Although Hero of Alexandria demonstrated a working reaction turbine in about 150 BC and Leonardo da Vinci had a basic knowledge of the power available in steam, it remained for Newcomen to develop the atmospheric engine in 1712. In this engine, a vertical cylinder—open at the top—was fitted with a piston suspended by a chain from a teeter-totter arrangement with a counterweight heavy enough to lift the piston. Steam at atmospheric pressure was allowed to flow into the cylinder and the steam valve then closed. A jet of cold water was then squirted into the cylinder, which cooled and condensed the steam. Atmospheric pressure then forced the piston down. These engines were slow, cumbersome, very inefficient, and relatively feeble. A huge engine with a 4-foot diameter piston and a 6-foot stroke operating at 10 strokes per minute would produce only about 3 hp.

James Watt realized that if the engine were made to operate by introducing the steam into the cylinder under some pressure greater than atmospheric and then allowed to expand, pushing out the piston, and finally allowed to condense and cool in another vessel, he could obtain a far more powerful and efficient engine. The early Watt engines reduced the steam consumption for a given amount of work by 75 percent compared to the Newcomen engines. The Watt engine was aided by the development of the machine boring mill by John Wilkinson, which made it possible to obtain much better fit between the piston and cylinder. By 1783, Watt had developed an engine capable of supplying rotary output. The steam engine very quickly became a commercial success since it was introduced at a time when virtually every available water power site in England was occupied and English industry had expanded to the point where cheaper and more abundant power was a virtual necessity.

It is noteworthy that our number for horsepower was developed by James Watt. Watt found that on an eight-hour shift, pumping water from a Welsh mine by walking on a treadmill, the horses averaged about 550 ft-lbs of work per second. The number was used to tell the potential buyer the number of treadmill horses the engine would replace.

The use of steam grew rapidly, and there were great inducements for inventors to improve the steam engine. By 1815, Oliver Evans of Philadelphia had built an engine operating at a pressure of 200 lb per in$^2$. This engine was used to power a self-propelled dredging machine that walked on land and paddled itself through the water. The machine was called the Oruckter Amphibolos. This machine was one of the very first self-powered vehicles.

While the development of the steam engine leaped ahead under the hands of practical men like Evans, another group of men were studying the basic processes of heat, expansion, and the behavior of gasses, liquids, and solids. Better understanding of these processes made the development of the efficient heat engine possible. Some of the principles must be grasped in order to understand the operation of a practical solar electrical plant.

# Charles's Law

Figure 11-3 shows a very simple experiment that can be performed at home. A glass tube of uniform bore is heated in a gas flame and the bottom sealed. While

$$V_t = V_o (1 + \beta t)$$
$$\beta = 1/273 \ 1° C$$

$V_o$ = Volume at 0° C

NOTE: The Kelvin temperature scale is equal to the Celsius temperature plus 273°.

**11-3**  Charles's law.

the bottom is still hot, the top is dipped into some water containing a bit of food coloring. The contraction of the cooling air inside the tube will cause a drop of water to be sucked up. This water drop forms a freely moving piston or seal on the tube, and the pressure inside the tube will be equal to the atmospheric pressure outside the tube. The volume of the gas inside the piston will be proportional to the height of the bubble.

If the tube is first dipped into ice water and the height of the bubble recorded, and then dipped into boiling water and a second height recorded, a graph like that on the right of the figure may be constructed. An interesting feature of this graph is the fact that the curve may be extrapolated to indicate that the volume would be zero at –273°C. Over the range from 0°C to 100°C the graph will prove to be quite linear, and the coefficient B will be found to be 1/273 to reasonable tolerances. Stated differently, the gas will expand 1/273 of its volume when heated from 0°C to 1°C and will contract the same amount, if cooled.

# Constant volume behavior

The apparatus shown in Fig. 11-4 is a bit more sophisticated. In this device, the gas is contained in a reservoir sealed in by a mercury column. The pressure within the reservoir may be determined by measuring the height of the mercury column above the surface of the mercury in the mercury reservoir. If the mercury reservoir

$$P_t = P_o (1 + \beta t)$$
$$\beta = 1/273\ 1°\ C$$
$$P_o = \text{Pressure at } 0°\ C.$$

**11-4**  Constant volume.

is much larger than the bore of the mercury tube, the volume of the gas is held constant to a first approximation.

The surprising thing about this experiment is the fact that the graph of the results tells us the same thing that the Charles's Law experiment did, but in a somewhat different way. An extrapolation of the results indicates that the pressure of the gas would be zero at –273°C. The constant B is also identical to that obtained from the Charles's Law experiment.

# Boyle's Law

Another one of the significant properties of gasses is enunciated in Boyle's Law. Figure 11-5 shows a cylinder and piston arrangement with a gas captive. Boyle's Law states that at a constant temperature, the product of pressure and volume is a constant. The curve of P/V is a hyperbola as shown in the accompanying graph.

In Fig. 11-6 we show something a little bit different. If we slightly relax the pressure on the piston, it will move outward to a new position. In doing this it has done some work since the force, which is equal to the area of the piston times the pressure in lb/in² moving through a distance *ds*, represents physical work. In order to keep the temperature from falling, enough heat must be added to the cylinder to replace the work given up to the mechanical forcing out of the piston. We note that the pressure is actually falling during this travel, but if we consider that the actual

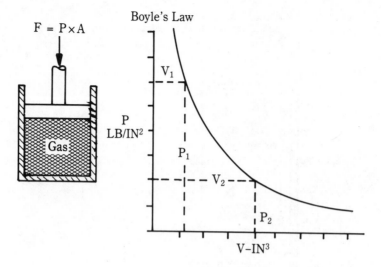

$$F = P \times A$$

Boyle's Law

$$P_1 V_1 = P_2 V_2$$

At a constant gas temperature

$$P_1 V_1 = P_2 V_2$$

**11-5** Boyle's law.

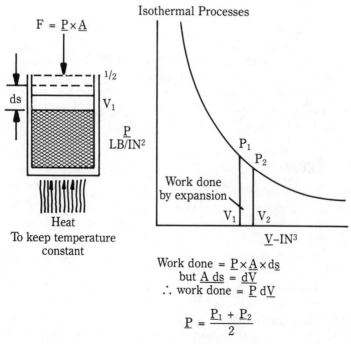

$$F = \underline{P} \times \underline{A}$$

Isothermal Processes

Work done by expansion

To keep temperature constant

$$\text{Work done} = \underline{P} \times \underline{A} \times d\underline{s}$$
$$\text{but } \underline{A}\, d\underline{s} = d\underline{V}$$
$$\therefore \text{ work done} = \underline{P}\, d\underline{V}$$

$$\underline{P} = \frac{\underline{P}_1 + \underline{P}_2}{2}$$

**11-6** Isothermal processes.

percentage of the expansion $dV$ is very small, then the amount of work that was done and the amount of heat that must be added is equal to $PdV$. This is called an *isothermal expansion* because of the constant temperature. If the piston had been force IN, the work would have been done on the gas and the cylinder would have had to give off an amount of heat equal to the mechanical work. As long as a perfect gas is maintained at a constant temperature, it will reasonably follow Boyle's Law for the pressure-volume relationship. Real gasses do not exactly follow the perfect gas laws. However, for our purposes here it is not worth complicating the picture.

# Adiabatic processes

When a gas expands or contracts without the addition of heat to it or the extraction of heat from it, the process is said to be *adiabatic*. This would happen if the cylinder and piston were perfect heat insulators so that no heat could enter or escape during expansion or compression. In actual practice, certain processes which take place rapidly can approach this performance. For example, the air escaping from a balloon or tire that was at the ambient temperature will be perceptibly cold since it was expanded and has not had time to acquire heat from the atmosphere. Similarly, the barrel of a tire pump or the cylinder of an air compressor will become very hot when in use because the mechanical work involved in the pumping is partially converted to heat, which does not have time to radiate off during the brief compression stroke.

In an adiabatic process, the equivalent of Boyle's Law is as follows:

$$P \, V^{\alpha} = \text{a constant}$$

The value of the exponent $\alpha$ is determined by the molecular structure of the gas. For most of the common gasses, such as hydrogen, oxygen, and nitrogen—which are diatomic—the value of $\alpha$ is 1.40. Some of the inert gasses, such as helium and argon, are monatomic and have a value for $\alpha$ of 1.66. Air composed mostly of diatomic gasses has a value of $\alpha = 1.40$.

Physically what this means is that in the rapid compression of a gas where the heat has little time to escape, the pressure winds up somewhat higher than it would if the compression took place slowly enough for the temperature to remain constant. Thus an adiabatic compression requires more work than an isothermal one. If the compressed gas is allowed to sit in storage for any period of time, it will come to equilibrium with the surroundings and the added heat will be lost.

Actual physical processes are never exactly isothermal or adiabatic but tend to lie somewhere between the two.

# The ideal heat engine

In about 1830, a French physicist, Sadi Carnot, described an ideal heat engine. The efficiency of this engine cannot be exceeded by any engine working within the same temperature limits. Actually, the Carnot engine cannot be built. However, it is useful to consider because it tells us exactly how much energy we could extract from a heat engine if all the processes were perfect. It is also useful for our purposes since it graphically illustrates the requirement for each of the major components of any heat engine.

Figure 11-7 shows the basic parts of the Carnot engine. A piston and cylinder that are perfect heat insulators are capped off by a cylinder head that is a perfect heat conductor. There are three stands on which the cylinder can be placed. The first is a stove or source of heat at temperature $T_1$. The second is an insulator that is incapable of transferring any heat. The third is a refrigerator that can extract heat from the cylinder at temperature $T_2$.

Starting with the cylinder standing upon the stove, the piston is held near the bottom of the cylinder. The cylinder contains a very small volume of gas at high pressure and at a temperature $T_1$ since the cylinder head transmits heat without restriction.

On the pressure versus volume diagram the unit is at position A. With a slight pressure reduction, the piston moves out slightly and the gas expands. This would tend to cool the gas. However, heat flows unimpeded through the cylinder heat maintaining the gas at temperature $T_2$. The gas expands along the isothermal curve AB, doing work by forcing out the piston. The gas absorbs a quantity of heat $Q_1$.

Next, the cylinder is moved to the insulating stand. On this stand the insulator prevents further heat transfer through the cylinder head and the gas undergoes a further expansion. This expansion is adiabatic, and the expansion is allowed to continue until the temperature falls to $T_2$. On the P versus V diagram the expansion follows the curve BC. During this expansion the piston is also delivering work to the outside world.

The cylinder is then placed on the refrigerator. Work must now be delivered from the outside world to force the piston in. The heat $Q_2$ that is generated flows freely through the cylinder head to the refrigerator; the compression is isothermal along path CD. This compression requires minimum work, and the gas remains at temperature $T_2$.

At point D, the cylinder is again placed on the insulating stand, and the piston absorbs work from the outside world to compress the gas back adiabatically to the volume and pressure of point A. The work delivered to the outside world is proportional to the area circumscribed by ABCca, whereas the work required to get the gas ready for the next cycle is proportional to the area circumscribed by CDAac. It can readily be seen that the net mechanical work delivered by the engine is proportional to area ABCD. The engine has delivered net work from the process and has transported heat from the stove to the refrigerator in the process. The Carnot Efficiency is given by the formula:

$$\text{Carnot Efficiency} = \frac{T_1 - T_2}{T_1} = \frac{Q_1 - Q_2}{Q_1}$$

where the temperatures are measured from absolute zero or −273°C (the temperature where the pressure and volume of a perfect gas would be zero).

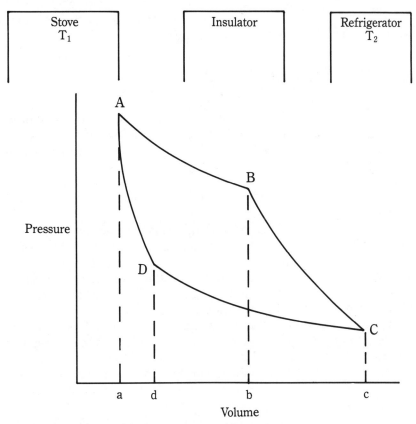

**11-7**   The ideal heat engine—the Carnot cycle.

The significance of this relationship is very great. Suppose we operate the Carnot engine with $T_1$ at the boiling point of water (100°C) and the refrigerator at a room temperature of 25°C. The Carnot Efficiency would be:

$$\text{Carnot Efficiency} = \frac{(100 + 273°C) - (25 + 273°C)}{(100 + 273°)}$$

$$= 0.201$$

At the very most, using the ideal engine we could recover only about 20 percent of the heat energy.

By comparison let us suppose that an ideal converter permitted us to operate the stove at the temperature of the sun or 6150°K while the refrigerator remained at room temperature. The Carnot Efficiency in this case would be 95.2 percent. The engine would be delivering about 4.74 times as much power for the same total heat (not temperature) input. Actually, there are no known materials from which such an engine could be constructed. As a practical matter, engines are limited to operating temperatures on the order of 500°C which is equivalent to 773°K or 932°F if the engines are continuously exposed to the hot gasses. With a room temperature refrigerator the Carnot Efficiency of such an engine would be 61.4 percent. This engine would extract more than 3 times as much power from the same quantity of heat as the low temperature example.

# The Carnot vapor cycle

The Carnot cycle thus described and the gas laws apply to engines operating with perfect gasses. We shall be discussing such heat engines shortly, and they can be made to operate on a cycle approaching the Carnot cycle. Engines such as the Sterling cycle engine are currently being investigated by several firms as alternate means of propulsion for an automobile.

By far the more common is the conventional steam or vapor cycle engine. Because of the differences in behavior the P versus V diagram of such an engine is considerably modified.

Consider an engine identical to the Carnot engine shown earlier, except that the cylinder is filled with a liquid at the boiling point at point A. Figure 11-8 shows that the addition of heat from the stove causes boiling under constant temperature and pressure conditions. At point B, the vaporization of the fluid is complete and the cylinder is transferred to the insulating stand. The vapor then adiabatically expands from B to C, while the temperature drops to $T_2$.

The cylinder is then moved to the refrigerator and the volume is reduced at constant temperature and pressure. At this point some of the heat of vaporization that was acquired to boil the liquid going from A to B is rejected to the refrigerator. This corresponds to path CD. The vapor is restored to liquid form. The cylinder is then moved to the insulator and the remaining small compression restores the condition to point A.

Both this vapor cycle and the gas cycle previously described are reversible. The vapor cycle actually closely approximates the performance of a real steam

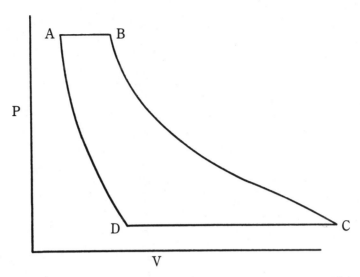

**11-8** The vapor cycle. The area under the curve represents the work done by the engine.

engine or household refrigeration unit. Because of the high heat of vaporization of the working fluid compared to a gas, the volume to be handled can be smaller in the vapor cycle engine than in the gas cycle engine.

# Summary

The nature of heat engines and the properties of gasses and vapors make it important to operate the engine at as high a temperature as is practical for the materials and techniques employed in the machine construction. For this reason, the construction of the solar electrical plant differs widely from the construction of a solar heating apparatus. In the latter, passive solar panels are frequently operated with a temperature rise of only 8° to 10°F. For this small rise an efficiency of only 3.2 percent would result, and the plant would have to be unacceptably large to produce any electricity. The power output would be only one-twentieth as great as it would be in a high-temperature system.

These factors add up to the fact that a practical solar electrical system must employ concentrators capable of producing the high temperatures required for reasonably efficient operation. The next chapter is devoted to a discussion of such concentrators and mechanisms for tracking the sun.

# 12
# Solar collectors

THE DISCUSSION OF THE PREVIOUS CHAPTER WAS INTENDED TO SHOW something of the geometry of the earth-sun situation and to emphasize the requirement for the development of the highest possible temperatures in the solar collector apparatus if the object is to operate a heat engine for the generation of electricity. In this chapter we shall examine the fundamentals of the design, construction, and testing of solar collectors suitable for the operation of heat engines. In order to understand the principles involved, we shall have to treat a few selected topics in the areas of optics, geometry, and electronic servo control. We shall attempt to go no further into these topics than is absolutely necessary to provide an understanding of the factors involved in the design of a workable solar collector.

## Optics

A solar collector has certain fundamental similarities to an optical or a radio telescope. One of the prime goals is the collection of the largest possible amount of power incident upon the *aperture* (or collecting area) of the instrument and directing it to a specific point or area where it can be used in some manner. The collector should also be equipped to track the body under consideration with minimum complexity so that the image is stationary with respect to the energy collecting medium.

There are also certain rather fundamental differences. Both the optical and the radio telescope are designed to capture the largest possible amount of the energy incident upon the instrument aperture and to provide the highest possible *resolution* (the ability to distinguish between two small adjacent bodies) so that closely spaced bodies can be separated. Both are measuring engines, rather than power collecting engines.

In order to provide the desired resolution, both the radio and the optical telescope must have *path length errors* (distance from object to reflector to collecting apparatus) that are small with respect to the wavelength. In the optical telescope, the wavelengths of visible light range from 0.72 to $0.38 \times 10^{-6}$ meters. For the radio telescope the usable range of wavelengths is from about $1.5 \times 10^{-2}$ meters to something on the order of 150 meters. The wavelengths on either side and the gap are effectively shut out from an energy standpoint by the nature of our atmosphere. From a practical standpoint, the energy available from moonlight and starlight are insignificant for the production of home electricity. Only the energy from the sun is of any practical significance. Again, as a practical matter, it is only the energy lying between 3 and $0.3 \times 10^{-6}$ meters that makes any significant contribution to the generation of household electricity. The shorter wavelengths are effectively absorbed by our atmosphere, and the longer wavelengths are too feeble to contribute any meaningful amount of heat to a heat engine.

Another factor that adds to the differences in design between solar collectors and telescopes is the fact that unlike stars, which have no measurable subtended angle, the sun has an appreciable subtended angle. The diameter of the sun measured from the earth subtends an angle of about a half a degree in the visible wavelengths and about 0.66 degree at the longer wavelength limit. The reason for the variability is that while the solar corona does not emit much visible light, it is still a copious emitter of longer infrared wavelengths and radio wavelengths.

In a solar collector, it is a particularly worthwhile objective, not to obtain a well-defined image of the sun, but simply to collect the largest reasonable amount of the energy. For this reason, it is not necessary to build the solar collector with the extreme precision of optics that is required for an optical telescope. As will be noted shortly, designers of solar collectors usually take advantage of this fact because of the economies involved.

There are three basic types of optical telescopes. The *refracting, reflecting,* and *catadioptric.* The refractor consists of a lens type *objective* (the light-gathering lens at the front of the telescope) and an eyepiece for viewing the image. The reflecting type makes use of a mirror for the objective. The catadioptric type employs a combination of lens and mirror for the objective to reduce spherical aberration and to reduce the tube length for a given magnification. All three types are used for solar collectors, although the catadioptric is generally employed only when available as surplus because of the expense. We shall discuss the refracting type first since this is the easiest type to understand and visualize. Before we do this, it is necessary to discuss a few topics that are necessary for the understanding of some of the behavior of the optical components.

# Angle of incidence = angle of reflection

Figure 12-1 illustrates the reflection law. It assumes that:

- Light travels at a fixed velocity in free space and at essentially the same velocity in air; on the diagram, $C_o$ represents that velocity.
- The propagation of light is in a direction normal to the wave fronts.

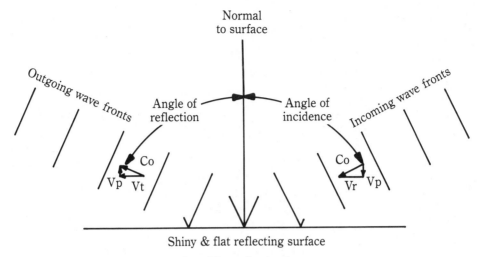

**12-1**   The reflection law.

The illustration shows a train of light waves traveling from right to left across the picture. The velocity of light in the direction of travel can be resolved into components tangent to the surface and perpendicular to the surface. These are noted as $V_t$ and $V_p$, respectively. If the light waves collide elastically with the surface, we see that $V_t$ will be unchanged, but $V_p$ will be reserved in sign. Thus the little triangles that constitute the resolution of $C_o$ will be mirror images of one another. Therefore the angles of incidence and reflection must be identical!!!

# Snell's law—refraction

Figure 12-2 shows a ray of light or a parallel wave train incident upon a piece of glass. The followiing formula was developed by a Dutch astronomer and mathematician, Willebrord Snell, in 1621 to describe the action and is usually referred to as Snell's law. It states simply that the sines of the angles of incidence and refraction are proportional to the ratios of the velocities of propagation in the two media.

If the wave fronts are pictured as a number of ranks of soldiers and the glass/ air interface is pictured as a junction between a muddy field and smooth pavement, the analogy explains the action. As the soldiers at one end of the rank encounter the mud, they are slowed and begin to fall behind the rest of the rank. The rank acquires a kink in the middle. When the soldiers all have finally entered the mud, the rank is again straight. However, the column has been bent to a new angle of travel and is now proceeding at a slower speed. The soldiers are also spread farther apart in the rank. Snell's law is useful in that it describes the operation of lenses and prisms and also shows some of the limitations on the design of these items.

One factor that is of interest is that since the sines of the angles can have only the values between zero and one (the negative values have no physical significance here), there is an angle beyond which the light cannot pass from the denser to the

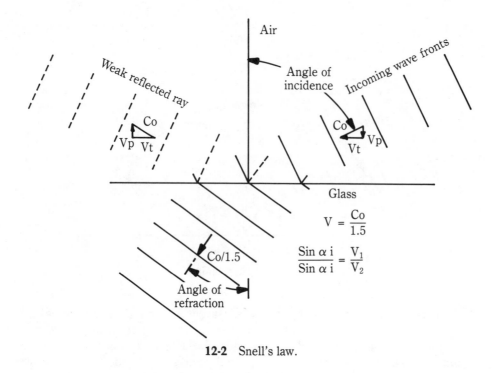

**12-2** Snell's law.

$$V = \frac{Co}{1.5}$$

$$\frac{Sin\ \alpha\ i}{Sin\ \alpha\ i} = \frac{V_1}{V_2}$$

less dense medium but is instead totally reflected. This angle is given by the formula:

$$\text{Angle of Extinction Formula:} \quad Sin\ \alpha\ i\ = \frac{V_2}{V_1}$$

For glass with an index of refraction of 1.5, this angle is 41.8°. No transmission is possible, at any larger angles of incidence. This is the effect that makes a 45-45-90° prism behave as if a mirror existed along the hypotenuse when the prism is viewed through one of the shorter faces.

In Fig. 12-2, you will note that a weak reflection was noted also. The reflection follows the angle of incidence/reflection law. Whenever light encounters a medium of different index of refraction, a certain percentage of the light is reflected. Sunlight is randomly polarized and can be thought of as having two components with polarization perpendicular to and parallel to the interface between the media. The amount of reflection is different for the two components and is a function of the ratio between the indices of refraction and the angle of incidence as well. The curve of Fig. 12-3 shows the variation of the reflection coefficients for the two components. It can be seen that the component polarized normal to the interface starts at normal incidence with a reflection coefficient of about 4 percent and actually drops to zero at an angle called the Brewster angle. It then rapidly climbs to unity. This property was discovered in 1816 by Sir David Brewster. At this angle 100 percent of the reflected light is polarized. All of the normally polarized component is transmitted without loss and only the parallel polarized component is reflected.

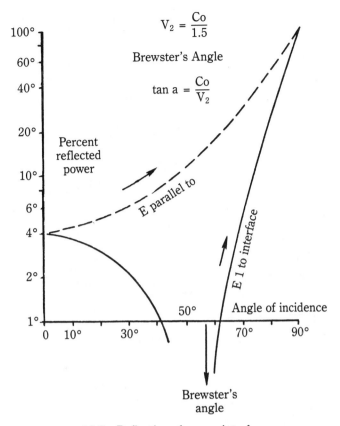

**12-3**  Reflections from an interface.

Beyond about 80°, the majority of the incident light is reflected and very little transmitted by the glass. These factors place certain limits upon the design of lenses and certain types of reflectors or mirrors.

# The simple lens

Figure 12-4 shows some of the components of a very simple plano-convex lens. If the assumption is made that the lens is very thin, then the focal length of the lens is given by the formula shown. For a lens with an index of refraction of 1.5, the rays of light from a distant object will be brought to a focus at a point equal to about twice the radius of curvature of the lens. As the parallel rays of starlight strike the curved front surface of the lens at increasing radius from the center, they do so at an increasing angle of incidence, thereby causing more refraction. The rays through this mechanism all come to a focus or point of concentration P.

Unfortunately, the thin lens is of little use for solar collection. As shown in the lower portion of the illustration, if the original object has some significant angular extent, the image is spread over about the same angle behind the lens, which

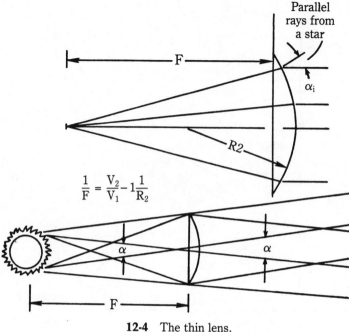

$$\frac{1}{F} = \frac{V_2}{V_1} - 1\frac{1}{R_2}$$

**12-4**   The thin lens.

makes it large and diffuse rather than compact and hot. The very assumption of a thin lens implies that the front spherical surface never makes a very large angle with the incoming parallel rays, so the focal length is large with respect to the lens diameter.

As the lens starts to become thick, the compensation between the angle of incidence and the point at which the rays strike the lens begins to break down and the lens no longer will function well to focus the rays. Under these circumstances it becomes necessary to make the lens aspheric or to make it curve less rapidly than a sphere does in the outer portions. Such lenses are actually made for use as condensing lenses in slide projectors and enlargers, but one that was large enough for solar collector use would be intolerably heavy. This problem can be solved by modifying the lens.

Figure 12-5 shows a modification technique. Suppose that you took a large glass hemisphere and cut the center of it out with a cooky cutter-shaped tool and then proceeded to remove progressively larger annular rings. If the center plug and the rings were then stacked on a flat surface with round surface down, then glued together, and the projecting back ground off, you would obtain a figure that would appear in cross section like the illustration. The majority of the weight would be

**12-5**   The Fresnel lens in section.

removed. However, the device would still have the properties of a thick lens, including the short focal length for maximum image concentration.

Because of the immense difficulties involved in making such a lens of optical quality, this type of lens has never been popular for astronomical or photographic work. However, the Fresnel lens has found wide popularity for lighthouses, searchlights, railroad lamps (and lately for magnification of TV images), where a very large, short, focal-length lens of not-too-great image clarity is required.

It is obvious that a certain fraction of the light in the cracks between sections will be lost. However, light-gathering efficiencies in excess of 80 percent are readily achievable. Fresnel lenses molded by a plastic replication process have found wide popularity as solar collectors. These lenses can easily be molded with aspheric correction to provide very large light-gathering power and very small focal length-to-diameter ratios for a hot, intense image.

The lens has an advantage over the typical reflector in that the lens need not be held with any great precision if image quality is not a prerequisite. The lens-type solar collector can therefore be built lighter and perhaps cheaper than a reflector type of equivalent size.

# The reflector type of solar collector

Figure 12-6 shows a section of a spherical surface with parallel rays of starlight incident upon it. For the rays striking the surface close to the axis it can be seen that the angle to a point halfway between the vertex $V$ and the center of curvature $C$ makes an angle very close to the angle between the radius of curvature and the incoming ray. In fact, the approximation is so good that it will yield astronomical quality images if the diameter of the reflector portion used is not greater than 1/10

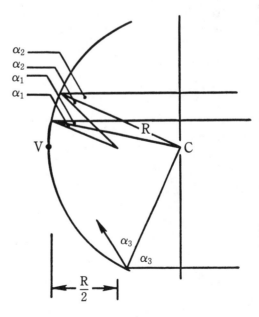

**12-6** The spherical mirror.

of the focal length or $1/20$ of the radius. However, in the outer portions of the sphere, the angles with the subscript "3" show that the quality of the focusing breaks down rapidly. This is termed *spherical aberration*. As with the thin lens, the long-focal-length reflector is of little value for solar collecting. Yet there is a relatively simple aspheric correction that can be made to the surface to eliminate this problem. Refer to the figure to see that the problem with the focusing from the outer portions of the reflector is due to the fact that the reflector curves around too fast. A parabola, which shows an increasing radius of curvature with distance from the vertex, can be shown to focus perfectly both the inner and the outer rays for a point source. In astronomical work, other factors such as image distortion tend to

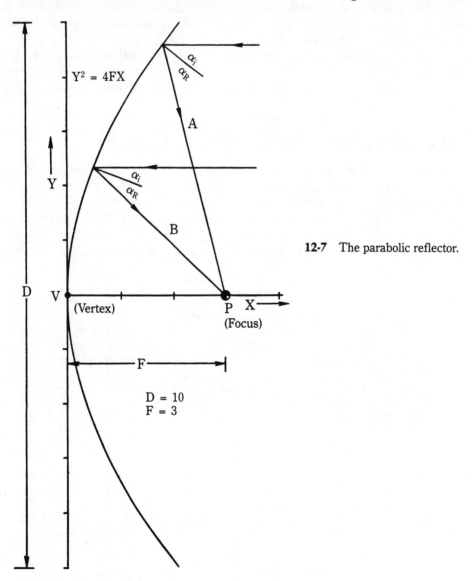

**12-7** The parabolic reflector.

limit the usable designs to a mirror diameter ¹/₆ or ¹/₈ of the focal length. But with radio telescopes and radar antennas, the use of diameter-to-focal-length ratios as high as four is common. Figure 12-7 shows such a parabolic reflector.

It is noteworthy in this design that the heat-collecting apparatus must be able to accept the focused energy over a wide range of angles. If the diameter of a reflector of this same focal length were carried to 12 units, the rays from the periphery of the reflector would be reaching the focus at right angles to the central ray. The parabola shown is still a very "fast" objective. To the photographer it would have an "f-number" of 3/10.

For photographic or astronomical telescope purposes, such a "fast" reflector would be of little use. The distance from the point of reflection to the focus for a point on the periphery ($Y = 5$, $X = 2.08$) is 5.08 units compared to the distance for a ray along the x axis of 3 units. This means that the image of a source of finite extent, such as the sun, would be larger by the ratio of 5.08/3 for the rays from the periphery compared to central rays. This would hopelessly blur the image. However, for applications such as a radar reflector or a solar collector where no attempt is made to achieve a well-defined image of the source, this does not represent a serious fault. The very short focal length provides a compact mechanical structure, which can be made very rigid.

Workable solar collectors can usually be made "faster" in the reflecting form than in the refracting form. This is due principally to the angle-doubling properties of the incidence/reflection law. Even in the relatively extreme design shown, the angle of incidence for rays parallel to the axis is only 50.19°. For a refractor with the same diameter and focal length, the rays at the periphery would have to have been bent through an angle of 59°, which is considerably more of a problem.

# Front and rear surface mirrors

In the discussion up to this point, it has been assumed that the reflection took place from the front surface of the mirror. Astronomical telescopes are usually constructed in this way to avoid the confusion of image shown in Fig. 12-8. The incoming ray will reflect some portion of the light from the first surface, refract into the mirror, reflect from the silvered (or aluminized) surface, and refract back out of the glass with some of the energy reflected back in. If the angle of the returning ray

**12-8** The rear-surface mirror.

Silver

Glass

exceeds the angle of extinction, the ray might be trapped within the glass. At the very least, the rear surface will give rise to multiple images. By itself, this is probably not too serious in a solar collector. However, the trapped rays will tend to concentrate toward the center of the mirror and contribute to heating of the glass. On a large solar collector, this heating could be sufficient to fracture the glass.

There is a distinct advantage to placing the reflecting surface on the rear side of the glass. This is the way that hand mirrors and rear-view mirrors are constructed because it protects the silvered surface from scratching. If the mirror gets covered with dirt or dust, it can be simply washed just as one would wash a plate glass window. By comparison, a front-surface mirror on glass requires great care in cleaning to avoid scratching or even wearing off of the reflecting surface. In addition, the rear-surface mirror has the advantage that the reflecting surface is protected from exposure to the atmosphere by the glass and is thus relatively immune to oxidation or tarnishing.

There are certain coatings that have been developed to protect the front-surface mirrors used in astronomical telescopes. These coatings are vacuum-deposited to a thickness of a few molecules and because of their extreme thinness do not disrupt the image formation. However, they would be extremely expensive to apply to a reflector as large as a solar collector.

After a discussion of the requirements of a solar collector, we shall return to the related discussion of first and second surface mirrors. At the present writing the lack of an inexpensive, durable, thin coating for large reflectors essentially precludes the use of first surface mirrors for practical solar energy collection. The surface would be too expensive to produce and too difficult to keep free from dust, dirt, and tarnish (or oxidation and weathering).

# The required solar image

With either a lens or a reflector system, even the finest optical element would reduce the image of the sun only to a disc at the focus with a subtended angle of approximately a half degree measured from either the center of the lens or the vertex of the parabola. If this were done, the image would attempt to assume a temperature equal to that of the sun or 6150°K, which is equivalent to 10,610°F. There is no known material that will remain solid at this temperature (or even liquid, for that matter). It is therefore necessary that the device receiving the image be cooled by some mechanism since the temperatures are high enough to cause firebrick to dribble and run like water. In actuality, it is not necessary in a solar collection system to have the image collected down to a focused image; it is necessary only to have enough concentration to obtain the desired heating of the working fluid for the heat engine.

**Note:** The shift over into English units is made here because the overwhelming portion of the engineering data available for boiler and steam engine design in the US is given in English units. The conversion factor for Kelvin temperatures to Fahrenheit temperatures is:

$$[ (°K - 273) \times 9/5] + 32 = °F$$

On the absolute temperature scale in English units the temperatures are given in degrees Rankine. This is 459.4° higher than the Fahrenheit temperature. Absolute zero degrees Rankine is equal to –459.4°F.

$$°R = °F + 459.4$$

Therefore:

$$[ (°K–273) \times 9/5] – 491.4 = °R$$

Most heat data are given in British thermal units (Btu). One Btu is the amount of heat required to raise the temperature of 1 lb of water by 1°F. The calorie is the metric unit. One calorie is the amount of heat required to raise the temperature of 1 gram of water 1°C. There are 252 calories in one Btu. Note also that the capital "Cal." used in food references is a kilogram calorie. It is 1000 times as large as the cal.

$$\text{One horsepower hour} = 2546 \text{ Btu}$$
$$\text{One kWh} = 3413 \text{ Btu}$$

It has been estimated by NASA that the highest direct flux of solar energy to be expected in the United States is 330 Btu/ft²/hr. By comparison a modern coal- or oil-fired boiler for steam power production will operate at a flux of 40,000 to 50,000 Btu per ft² per hr. If we take the boiler limit as being a reasonable engineering limit, we see that a concentration ratio of 150 is about what we would want to use for this maximum solar flux. If the installation were to be used in a northern climate and it was desired to operate with a feebler solar input at maximum efficiency, we could use higher concentration ratios. However, this would require some aperture reducing control to keep the solar flux on the boiler or heat gatherer within safe limits during the summer months.

## Size of collector

It is instructive to consider the size of reflector that we might require in order to produce a reasonable amount of home electricity. For our model system let us suppose that we wish to have the system operate at a solar flux level of 165 Btu per ft² per hr or one-half of the maximum expected. For a 1-kW output with an overall efficiency of 0.06, we would have to use a collector with an area of:

$$\frac{3413 \text{ Btu/kWh}}{0.06} = 165 \text{ Btu/ft}^2/\text{hr} \times \text{Area (ft}^2)$$
$$\text{Area} = 344.75 \text{ ft}^2$$

If the collector were round it would have a diameter of 21 feet. For our 3-hp equivalent, we would require 771.5 ft² and a diameter of 31.3 feet.

The 3-hp model is selected because even under the conditions of the equator, the solar plant will operate effectively for only a portion of the day and the need therefore exists to produce energy over and above the immediate usage for storage. If we only required that the unit produce the 2.24 kW (3 hp) at the peak solar energy level, the size of the collector could be shrunk to a 15-foot diameter. This is a much easier and less expensive size. However, it would only here and there and now and then meet the requirements for home electricity—even on the sunny Arizona desert.

The design and construction of a 32-foot paraboloid of revolution presents some problems, but reflectors of this size are commonly constructed for radar, satellite terminals, and radio telescopes. These units are also mounted on tracking pedestals or mounts. The principal obstacles are therefore not insuperable engineering problems but rather the economic constraints.

The collection efficiency figure employed in the calculations might seem excessively pessimistic at first glance. However, it will be noted later that it is not. The figure was derived as follows:

| | |
|---|---|
| Engine thermal efficiency | 0.089 |
| Engine mechanical efficiency | 0.71 |
| Boiler efficiency | 0.974 |
| Collector efficiency | 0.974 |

The overall product of these is the 0.06 number used in the calculations. The engine mechanical efficiency is intended to include the friction losses, departure from ideal cycle, inefficiency in the generator, and the work done on the auxiliary pumping and tracking machinery. In a small plant, such as the unit under discussion, an efficiency of this order is quite high.

The boiler efficiency number is intended to include the heat losses between the boiler and the remaining steam system. The collector efficiency number is intended to include the amount of sunlight lost in heat in the collector itself and the amount that is not absorbed by the boiler, as well as a certain amount of scattering due to dirt or dust on the collector. Here again the efficiency cited is quite high.

It is obvious that an increase in these numbers would make it possible to operate the system with a smaller collector. Yet if all the inefficiencies beyond the basic thermal efficiency were made equal to unity, the collector area would only be reduced by a factor of 1.48 and the collector diameter would be reduced by a factor of 1.22. The cost of collectors tends to run at something like the 1.3 power of the area. Therefore, the cost would fall by a factor of about 1.7 if the same quality of construction were able to provide the requisite improvement.

The purpose of this discussion is, of course, to illustrate the fact that it is probably economically warranted to go to considerable pains to obtain a relatively high efficiency in each of these departments. As with all engineering efforts, there must certainly be some point of diminishing economic return where the incremental improvement in performance is not balanced by the increased investment required. The history of solar steam installations is so limited that no rule-of-thumb guidelines are available, and the investigator is pretty much on his own to develop an economically feasible model.

# The hemispherical boiler

Both of the collectors discussed thus far have been figures of revolution about the focal axis. If the boiler is to be operated at the 50,000-Btu-per-ft$^2$-per-hr rate, it is apparent that the energy incident upon the collector cannot be focused down into too tight a spot. It can be shown that a hemispherical target concentric with the focus of the reflector will have a nearly uniform heat input over the area illuminated.

For a reflector such as the one illustrated in Fig. 12-7, with a diameter of 10 units and a focal length of 3 units, the subtended angle of the reflector seen from the focus is a cone with a half angle of 79.6°. Thus, the boiler should have the shape of a portion of a sphere with a subtended angle somewhat larger, perhaps 85°. The illustration of Fig. 12-9 shows a suggested arrangement in cross section. The water enters along the axis of the hemisphere and is constrained to flow between the dome-shaped surface receiving the solar energy and the backing

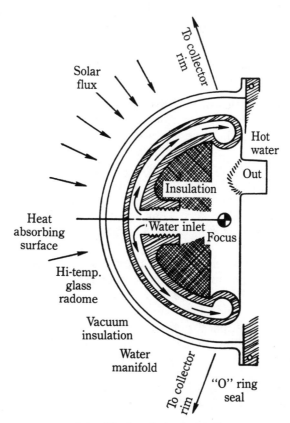

**12-9**   The hemispherical boiler.

plate. The water flows to the circular manifold and then out the outlet pipe. If the front of the dome is relatively thin, there will be a relatively small temperature drop between the absorbing surface and the water surface.

An example here would be instructive. Let us consider a solar collector of 2.24-kW electric output rating operating in an incident flux of 165 Btu per ft$^2$ per hr. For conservatism, we shall neglect the reflector losses. The boiler will be exposed to 130,600 Btu per hour. For our nominal 50,000-Btu-per-ft$^2$-per-hr boiler design we see that we would require a boiler collecting area of 2.612 ft$^2$.

The formula for the area of a segment of a sphere cut out by a cone is:

$$A = 2\pi r^2 (1 - \cos \theta)$$

For our reflector, $\theta = 79.6°$. Using this formula, we find the radius of the boiler to be 0.71 feet or 8.55 inches.

The design of boilers is generally done under ASME design rules, which have two factors of particular interest to us here.

- A boiler should be hydrostatically tested at 150 percent of the maximum rated pressure, struck a number of blows with a calibrated hammer, and then hydrostatically tested to 200 percent of rated pressure.

- Except in special cases, the maximum tensile strength of steel is taken as 50,000 lb per in$^2$.

Now, let us suppose that we wish to operate our boiler at 200 pounds per square inch gauge (psig). Because of the vacuum drawn on the outside of the boiler for insulation purposes, we would have a pressure differential of 215 pounds per square inch absolute (psia). To allow a bit of leeway, let us design the boiler for 450 psia. The projected area of a hemisphere of 0.71-foot radius is 1.594 ft$^2$, and the circumference is 4.475 feet. Multiplying the area by the pressure, we find that the boiler dome is holding a load of 49,341 lb! Dividing this by the periphery of the boiler, the force works out to 918.8 lb/per lineal inch. For a stress of 50,000 lb per in$^2$, we find that the boiler dome could be made of steel 0.018 inch thick (about $1/64$ of an inch), or #24 B&S gauge.

The conductivity of steel is about 26 Btu per hr per °F per ft thickness. At our 50,000-Btu-per-ft$^2$-per-hr input rate, the temperature rise across the heated portions of the boiler works out to 2.95°F. There would of course be a gradient brought about by the amount of temperature rise the water experienced in the flow from the center to the periphery of the boiler. This will be discussed in more detail later. However, it is worthy of note here that it is inadvisable to let the temperature gradient rise too high since this will cause warping and explosive stresses upon the boiler as well.

The message to be conveyed by the preceding exercise is the fact that there are tremendous forces involved in the solar boiler. The design of such a unit should not be undertaken by anyone not thoroughly versed in mechanical engineering practices. The American Society of Mechanical Engineers Boiler Code should be followed wherever local codes and ordinances do not override it. Unless properly designed, installed, and tested, boilers are *dangerous*. Boiler expansion used to take a toll regularly in human life and property!

# Insulation

Of course, the rear surface of the boiler can be insulated from heat loss to the atmosphere by any of the common high-temperature insulating materials such as fiberglass, ceramic foams, and so on. Asbestos used to serve this purpose. However, the dome itself would lose a great deal of the hard-won heat to the atmosphere if it were exposed. Unlike the back side, the dome must have free access to the solar flux being concentrated by the collector. This is probably most easily accomplished with a glass or pyrex transparent radome, which is evacuated to minimize convection losses to the atmosphere. At the operating temperatures of a covered boiler, the losses will be almost entirely due to conduction and, in an uncovered boiler, to convection currents and wind currents in the atmosphere. Evacuation of the radome will reduce these by more than an order of magnitude.

# Tolerances

One of the major factors affecting the practicality of a reflector-type solar collector is the question of the tolerances permissible in the construction. The reflector sizes thus far discussed are noticeably larger than the very largest optical telescopes. In order to maintain the image quality of these telescopes, it is necessary to maintain the "figure" of the mirror within a fraction of a wavelength of the design surface (which is not always a parabola). For this reason, one finds that the *cost* of these large telescopes, as well as the purpose, is astronomical.

By comparison, the solar collector does not have the requirement that the image be well defined. It is only necessary that the overwhelming majority of the energy strike the boiler. This point is important since even a fraction of the energy collected, if focused upon one of the boiler supports, would rapidly overheat that portion of the structure. If the situation were severe, it could melt the structure.

To examine this requirement, let us refer to Fig. 12-10. This illustration shows an ideal theoretical paraboloid with a flat mirror tile placed tangent to the paraboloid at the center of the mirror surface. The reflection of the sun from this mirror surface would spread about a quarter of a degree from the centerline. If we assume a paraboloid of 32-foot diameter with a focal length of 9.60 feet and a boiler radius of 0.44 foot, we would find that the image of the sun would spread about 0.91 foot on each side of the centerline. If the mirror were perfectly aimed so that the center of the image of the sun lay on the radius connecting the boiler center (the focus of the paraboloid) and the center of the mirror tile, we would find that it was possible for the mirror to have a diameter of 0.258 foot and still keep the entire reflected image of the sun within a 30° cone on the boiler dome.

Next let us suppose that we allow the mirror tile to tilt to the extent that the image of the sun is just tangent to the sphere described by the boiler. For the tile at the $Y = 16$-foot point, we would find that the ray from the mirror had deviated through an angle of:

$$\frac{0.44 - 0.91 - 0.258/2}{20.83} = \quad \text{arcsin } 0 = 0.61°$$

**12-10**   An ideal theoretical paraboloid with a flat mirror tile placed tangent to the paraboloid at the center of the mirror surface.

From the law of incidence and reflection we know that the angle through which the ray deviates is twice the angle through which the mirror is turned. If we divide the angle by two and multiply the sine of the angle by the diameter of the mirror tile, we obtain the height of the tilt, which is 0.00137 foot or 0.0165 inch.

If we try the same procedure with another mirror of half the size, we obtain a ray tilt of 0.783° and a mirror movement of 0.00088 foot or 0.0106 inch.

These numbers give us a feel for the required tolerances upon the reflector. The surface should not deviate from the surface of the true paraboloid by more than about 1/64 inch over a span of about 3.5 inches and by no more than about 0.01 inch in a span of about 1³/₄ inches. For a 32-foot structure, these are very tight tolerances indeed. However, I can attest from personal experience to the fact that it is possible to build a parabolic reflector of this size, with these tolerances. For a radar antenna operating at 6000 MHz, it is necessary to hold to such tolerances in order to attain a −30 Db sidelobe level. A number of antennas of this size with such specifications have been built in production runs for satellite communications, and radar usage. They are not very cheap.

Note that the deviations apply to a best-fit paraboloid. Provision is usually made to permit slight adjustments of the feed location on radar antennas for best performance, and a similar adjustment would be a wise provision for a solar collector. For a collector of this size an adjustment span for the location of the boiler of about 0.66 inch should be adequate if the unit is built with the tolerances above.

The actual use of flat mirror tiles is not at all impractical for a solar collector. The tiles could be laid upon a male mold of high precision and attached to a relatively imprecise paraboloid backing structure with a thixotropic (sticky, not runny) epoxy glue, such as car body compound used for repairing rust holes and dents. This technique of using a relatively imprecise backing frame and attaching it to a reflecting surface that is held to shape by a precise mold is frequently used in antenna fabrication. Since the mold can be used to make a great number of reflectors, it is economically advantageous to spend a substantial sum on the construction of the mold since this forms the surface contour for the reflectors to be produced. The back frames can then be made relatively imprecise at a considerable saving in construction cost. Obviously, this philosophy applies only when there is some volume production.

The use of the flat mirrors has another advantage in the fact that many of the objections to second surface mirrors are reduced or minimized if the mirror is not being used to focus a precise and well-defined image. There is also the obvious economic advantage to purchasing a ready-made item. While the aperture area of our 32-ft reflector is about 804 ft$^2$, the area of the surface of the reflector is about 869 ft$^2$. If good quality plate glass mirror tiles can be purchased for about $3 per ft$^2$, the reflecting surface can be covered for about $2,600.

The possibility exists of obtaining surplus paraboloidal antennas of the proper size. For radar work, these antennas are usually covered with a metal mesh or grating for the reflecting surface. In the absence of a precise mold, the surface of the reflector could be tiled by hand with the 16,600 or so tiles, each of which could be adjusted optically for focus when bedded in the cement. This would add about 3000 lb to the weight of the reflector if 1/4-inch plate glass mirrors were used. The increased wind resistance and the increased weight would probably still lie within the design limits of the reflector structure and the antenna mount. After all, these units are often designed to withstand 120-mph winds while covered with 1 inch of radial ice, which would generally choke the holes in the reflector surface. The overall weight of the antenna and mount thus treated will probably lie in the 5,000- to 6,000-lb range.

This solution is far less expensive than trying to build the reflector and mount from scratch. The reflector would tend to look a bit like the ball hung in the Aragon Ballroom but it would be quite effective.

# Alternate forms

A somewhat less difficult reflector to build is the parabolic cylinder-type collector illustrated in Fig. 12-11. This kind of reflector is parabolic in one direction and a straight line in the other. Because of this it focuses the rays of the sun into a line rather than a point. The boiler, therefore, can be constructed of tubing.

**12-11**   The parabolic cylinder collector.

The parabolic cylinder is easier to build than a paraboloid of revolution for rather obvious reasons. It has the advantage that it needs aiming precisely in only one plane since a slight mis-aiming in the linear plane only tends to move the image endwise along the boiler.

If one pursues the same calculation performed for the parabolic cylinder of 32-ft width 24-ft height, one finds that the 50,000-Btu-per-ft$^2$-per-hr rate would give a boiler tube only about 0.94 inch in diameter. The tolerances required for such a small target would be impossible at practical cost in a structure of this size. However, if the heat input rate were reduced to 2,500 Btu per ft$^2$ per hr, one would find that a boiler radius of 9.4 inches would result. The inset in the illustration depicts a boiler made of a series of standard tubes brazed together in a semicircle, enclosed in the glass tube radome. The reduction in heating rate would be accompanied by a similar reduction in flow rate through the boiler, if the same system pressure was to be maintained.

The low heating rate and the length of the boiler have some other problems in that a very substantial gradient in water temperature would exist if the flow were simply a straight line. The water at the exit end would be a great deal hotter than the water at the entering end. Some help could be obtained by making the water in the even-numbered tubes enter at one end of the boiler and the water in the odd-numbered tubes enter at the opposite end. The boiler would probably have to be segmented as well so that the water would enter and exit at a number of points along the structure.

A low transfer rate boiler has certain problems. In order to operate smoothly, a boiler should have the water enter and leave in water form. It should not flash off into steam inside the boiler tubes themselves. In a high transfer rate boiler, the high velocity of the water promotes turbulent flow through the tubes. This tends to scrub the tube walls and prevents the formation of steam films and bubbles. Since the specific heat of the steam is less than that of the water, these can give rise to local overheating of the tubes. The violent collapse of the bubbles can make the boiler noisy and can promote vibrations that fatigue the tubes. The use of the small tubes promotes turbulent flow, which tends to minimize this action. Another advantage to the use of the small tubes is the fact that for a given pressure, the wall thickness of the tube might be decreased by about the ratio of the diameter. This means that there is less metal through which the heat must flow, and the unit therefore responds faster.

An advantage to the parabolic cylinder collector tiled with plate glass mirrors is the fact that there are fewer tiles to be installed and adjusted since the tiles can be very long in the straight-line plane. The arrangement shown in Fig. 12-12 is a technique for aligning the tiles. If the angle $\theta$ is calculated along with the distance,

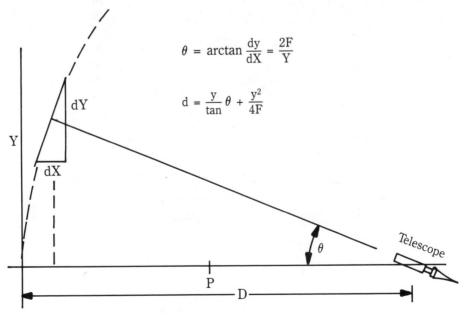

$$\theta = \arctan \frac{dy}{dX} = \frac{2F}{Y}$$

$$d = \frac{y}{\tan}\theta + \frac{y^2}{4F}$$

**12-12**  Tile alignment.

a surveyor's transit may be set up along the focal axis of the reflector. The tile is correctly aimed when the image in the mirror shows the transit precisely centered on its own cross hairs. This technique may also be used on paraboloids of revolution, in which case the angle might have to be resolved into azimuth and elevation components for tiles not in the principal plains.

# The power tower

It was noted earlier that the cost of reflectors tends to rise faster than linearly with area. As a matter of fact, the cost seems to correlate with about a 1.3 power of area. This is due to the fact that as the reflector gets larger in area, it becomes higher and requires even more support and structure to hold its shape in wind and under the pull of gravity. Since the 2.24 kW plant is seen to be noncompetitive with commercially available steam or oil-fired units on a cost per installed kVA basis, a

**12-13**   The Department of Energy's Central Receiver Test Facility at Sandia is the site of the country's first solar central receiver project to use molten salt in an electrical power generating system. Molten salt flows through a receiver located on top of the 200-foot-tall tower. After being heated in the receiver, the 940°F salt transfers solar energy to a water supply, thus making steam to drive a turboelectric generator. More than 200 mirror arrays, comprising 88,800 square feet of reflecting surface, direct the sun's energy to the receiver.

**Table 12-1. Molten salt electrical experiment.**

| | |
|---|---|
| *Receiver* | Inlet Temperature: 590°F |
| | Outlet Temperature: 1050°F |
| | Salt Flow Rate: 97,000 lb/hr |
| | Thermal Capacity: 5 $MW_t$ |
| | Average flux: 230 $kW/m^2$ |
| | Peak flux: 530 $kW/m^2$ |
| | Receiver Tubes: Incoloy 800, 3/4-in O.D. |
| | 18 serpentine passes, 16 tubes per pass |
| | Total area 18 ft wide × 13 ft high = 234 $ft^2$ |
| *Thermal Storage* | Hot Tank: 1050°F |
| | Corrugated liner .050-in Incoloy 800, |
| | 13-in firebrick insulation, .25-in carbon |
| | steel shell, 10 ft diameter, 23 ft high |
| | Cold Tank: 590°F |
| | Carbon steel tank with external insulation |
| | 12 ft diameter, 12 ft high |
| | Salt Capacity: 79,000 kg; 175,000 lb; 1420 $ft^3$; 10,600 gal |
| *Steam Generator* | Thermal Capacity: 3.1 $MW_t$ |
| | Steam Flow: 11,000 lb/hr at 950°F, 1100 psi |
| | Salt Flow: 65,000 lb/hr hot salt; 14,000 lb/hr cold salt |
| | Evaporator: Low-alloy chrome molydenum steel |
| | U-tube, U-shell, 2.15 $MW_t$ |
| | Superheater: 304 stainless steel |
| | U-tube, U-shell, 0.98 $MW_t$ |
| | Steam Drum Conditions: 567°F, 1200 psi |
| | Recirculation Ratio: 3.7:1 |
| *Turbine-Generator* | Power Rating: 750 kWe with 0.8 pf, 440v, 3 phase ac |
| | Steam Flow: 7,800 lb/hr at 940°F, 1050 psi |
| | Condensor: 5-in Hg (2.4 psia), 133°F |
| *Pipes and Valves* | Riser: 3-in carbon steel |
| | Downcomer: 2-in 316L stainless steel |
| | Hot salt valves: 316 stainless steel with bellows seals |
| | Cold salt valves: Carbon steel or stainless steel bodies |
| | with stainless steel trim and bellows |
| | seals |
| *Salt* | Composition: 60% Sodium Nitrate ($NaNO_3$) |
| | 40% Potassium Nitrate ($KNO_3$) |
| | Density: 113 $lb/ft^3$; 1809 $kg/m^3$ |
| | Heat Capacity: 0.365 Btu/lb-°F; 1532 W-sec/kg-°C |
| | Melting/Freezing: 429 to 473°F; 221 to 245°C |
| | Thermal Conductivity: 0.30 Btu/hr-ft-°F; 0.52 W/m-°C |
| | Heat Fusion: 46.8 Btu/lb; 108,900 W-sec/kg |

1 megawatt commercial unit (446 times the power) would be nearly 2800 times as bad on a cost basis. It is obvious that some other technique must be found for solar energy collection if it is to succeed on a commercial scale.

One possibility has been suggested in the form of the *power tower*. Figure 12-13 shows this concept. A central tall tower supporting a solar boiler is mounted in the middle of a field with an array of small to medium sized mirrors which are individually mounted ringing the northern side of the tower or perhaps completely surrounding it in low latitudes. Each of these mirrors is individually gimbaled so that it can send its reflected rays to the boiler.

This technique has been pioneered by the Sandia National Laboratories. The illustration in Fig. 12-13 is the 5 MW (thermal) test facility constructed at Albuquerque, NM. [**Note:** One MW = 1,000,000 watts.] This facility pioneered several things besides the construction of the collector. It makes use of a molten salt of 60% sodium nitrate and 40% potassium nitrate as an energy storage medium. The salt is heated from an entry temperature of 590°F to 1050°F at exit. Steam is made by passing the hot molten salt through a heat exchanger where it heats water for the steam system. The data of Table 12-1 describe some of the details of the system. The unit is capable of supplying 750 kW of electricity.

One of the advantages of this system is that the hot molten salt may be stored so that the electrical output can remain stable when the sun goes behind a cloud, and other such changes occur. Of course this comes at the penalty of having to supply enough initially to melt the salt so that it can be pumped through the system.

A larger 10 MW (electrical) unit has been constructed and operated to provide a limited amount of commercial electricity in the Barstow, CA region.

# 13

# Mounting and tracking of solar collectors

FROM THE PREVIOUS DISCUSSION, IT SHOULD BE APPARENT THAT IT WILL be necessary to cause the solar collector for an electric plant to track the sun in its motion across the sky. It is true that solar heating systems that do not track the sun are feasible. However, a solar steam plant requires a much greater concentration of heat than a solar heating system. For the latter, the only slightly concentrated solar flux is sufficient to warm the water. For the solar steam plant, though, the water must be not only warmed but boiled, which requires a concentration factor of at least 50 of the solar flux. The concentrator must be moved in order to have the concentrated flux impinge upon the boiler. The relatively high concentrations discussed in the previous chapter will generally require a tracking of the position of the sun to an accuracy on the order of a quarter of a degree.

The parallel between the solar collector and the optical or radio telescope was drawn earlier. A large number of different mountings have been developed for radio telescopes. The radio telescope becomes more nearly applicable because of the size to which the instruments are built and because of the usually short focal length of the instruments. By comparison, the optical telescope has a very long focal length for a given size and this reflects itself in the mechanical design of the mounting.

For a great many years, astronomical instruments were constructed mainly with *equatorial* mountings. In the equatorial mounting, one axis of the instrument was aligned very precisely parallel to the axis of the earth. Rotation of the instrument about this axis was said to be in the *hour angle*. A second axis perpendicular to this was called the *declination* axis. Once a star was located in both declination and hour angle, the declination axis could be locked and the star tracked by motion of the instrument in hour angle only at the rate of 15° per sidereal hour. The equatorial mounted instrument has several distinct advantages:

- The mechanical drive system need be applied to only one axis of the mount.
- The position of the object can be read by use of setting circles applied to both axes with no mathematical correction. The declination axis reads declination directly and the hour angle axis reads hour angle.

The principal drawback to the use of the equatorial mounting is that one of the shafts must be inclined with respect to the vertical for installations in all locations except for the north and south poles. Machinery as large as is required for a solar collector is much less expensive to build and install if the axes of the machine are either vertical or horizontal. For one thing, with an inclined axis machine, if there is any unbalance in any of the machinery it will tend to swing down to the lowest position. The drive for the machine will have to push the machine uphill until the equilibrium point is passed and will then have to hang on to prevent it from coasting downhill of its own accord.

By contrast, the *altazimuth* mount has one shaft vertical and the second shaft horizontal. The machinery is required to have a precise balance only about the *altitude* axis, and a small unbalance in the vertical *azimuth* axis will not encourage any self rotation effects.

Figure 13-1 shows a comparison of the two mounting types. It is usually desirable to have the two axes cross. In the case of the equatorial mounting, this means that the center of gravity of the collector and the declination drive does not pass through the center of the hour angle bearing. Thus, there is a considerable overturning moment on this bearing. By comparison, the altazimuth mount has the center of gravity of the rotated assembly directly on the axis of the azimuth bear-

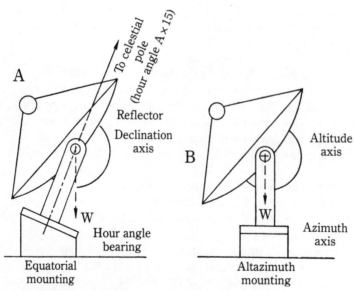

**13-1**   A comparison of the two types of mounting for solar collectors, the equatorial at A and the altazimuth at B.

ing. At low latitudes, the fork of the equatorial mounting can be extended, and a second bearing placed upon the top of a pier at the northern end of the hour angle axis. However, at midlatitudes, the altazimuth mount has a distinct structural advantage.

The principal drawbacks to the use of the altazimuth mount was that all the angles had to be calculated from the positions of the azimuth and elevation circles, the drive motion had to be applied simultaneously to both axes, and the drive was nonuniform with time. Since the development of the minicomputer, many of these objections have disappeared, because the minicomputer can be programmed to give a running translation of the declination and hour angle commands into azimuth and elevation angle commands with any required degree of precision as shown in Program 11-1. Accordingly, the use of the altazimuth type mount has become increasingly popular for astronomical instruments. The savings in the mechanical construction more than offset the costs of computerizing the drive.

Probably the cheapest way to obtain a suitable azimuth mount for a solar collector is through the surplus market. Most tracking-type radars employ a mount of this type and are directly adaptable to this application. Another possibility lies in the use of a five-inch naval gun mount. In the early days of radio astronomy during the 1950s, the five-inch gun mount was the most popular choice for a radio telescope mounting. A number of these gun mounts are probably still available from scrapped WWII destroyers and cruisers. The mount is extremely strong and stable and quite capable of supporting reflectors up to 50 feet in diameter in wind and ice.

# A home-brew mount

If one of the sophisticated antennae or gun mounts is not available through the surplus route, a suitable mount can be constructed of wood and available hardware along the basic pattern shown in Fig. 13-2. In this case we see an octagonal frame with a series of four trailer wheels mounted in a circle. A surplus truck rear axle half is used as a center pivot, but all the weight is carried by the trailer wheels. The wheels simply roll in a circle upon a cast concrete pad, and the truck axle is imbedded in the center of the pad. The main frames would probably look neater if they were mortised to lie in the same plane. However, this tends to reduce the strength by about one-half and the stiffness by a factor of four. Therefore, it is probably better to use them uncut. The design carries the weight around the periphery, and all the overturning moments are carried directly to the trailer wheels.

It would probably be good construction practice first to cast the circular pad for the wheels to run on, leaving an open excavation for the truck axle. The truck axle could then be attached to the finished turntable, leaving plenty of shim room to make up for variation in the tire pressure. Once the table was cemented on the track, the axle could be centered in place, thereby assuring that there was no interference with the truck axle. After the cement hardens, the axle can be loosened from the steel plate and the shims removed so that there is about an inch of vertical "float" and the axle carries no vertical load. If this is not done, the turntable will

**13-2**  A simple homebrew azimuth turntable.

be almost certain to bind in some positions and for some conditions of tire inflation. If the steel plate is attached from the top, the servicing of the axle will be
facilitated, and it will be possible to jack up one wheel for removal in case of a flat
tire.

The view of Fig. 13-3 indicates the fork and trunnion construction. This unit
relies upon an "A" frame type of construction to carry the weight and overturning
moment down to the trailer wheels rather directly. The sway braces are a very necessary part of the construction since the unit will want to be turned toward the sun
and the wind might approach from any angle.

Wind force upon this structure can be considerable, as noted in the section on
windmills. If we consider the device as having the same resistance as a flat plate,
the wind will deliver a pressure of:

$$P \quad = 0.0042 \ Va^2 \ lb/ft^2$$
$$V_a \ = \text{wind velocity in mph}$$

In a 90-mph wind, this force is 20.1 lb per ft$^2$. Considering only the area of a 32-foot
collector, this amounts to 16,165 lb. If we assume a height for the elevation trunnion of 20 feet, this amounts to an overturning moment of 323,300 ft-lb.

If the radius of the turntable out to the tires is 12 feet, and the weight of the
turntable and collector is assumed to be 8000 lb, the deadweight at the center of
the table will supply a righting moment of 96,000 ft-lb. If a ballast of 4800 lb were

**13-3** The elevation axis.

applied at each of the wheels, the unit would be in equilibrium. However, the ballast and deadweight would bring the wheel loads up to 6800 lb per wheel.

A far more practical approach would be to consider that the unit would be stowed with the reflector pointed vertically at all winds above some more modest level. If we were to design for operation in winds up to 40 mph, the maximum pressure would be 4.7 lb per ft$^2$, and the maximum overturning moment would be 3780 ft-lb. This is well within the deadweight righting moment. For higher winds the collector would be stowed pointing vertically to minimize the surface presented to the wind and the reflector itself could be guyed to the earth using prebuilt anchors installed at erection time. In northern climates, the reflector might have to be stowed pointed horizontally to minimize snow loads.

Note that the sketches in the illustrations have left out a number of items, such as gussets and bearings to captivate the elevation trunnion, for the sake of clarity. These items *cannot* be left out in the finished unit. The dimensions of the turntable have also been left out since these will be a function of the collector size. A device of this sort will generally be quite stable if the radius to the tires is about 3/4 of the radius of the reflector.

# Range of angles

The construction of the antenna mount is obviously influenced by the range of angles through which the reflector must be swung. Figure 13-4 shows the situation

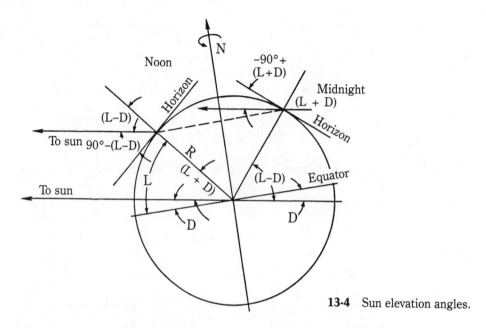

**13-4**   Sun elevation angles.

for an operation in the northern hemisphere. The sun is always highest at local noon (standard time) and the elevation angle above the local horizon is $90°-L + D$, with $L$ being the north latitude of the site and $D$ being the declination of the sun at the time. Actually, for reasons of stowage noted earlier, it is usually advisable to be able to point the collector straight up. The limitation is therefore placed by the stowage condition rather than the range of sun angles.

Although Program 11-1 gives a detailed mechanism for computing the azimuth and height of the sun, the following discussion will suffice for estimates of mount design.

The range of azimuth angles is also limited. The azimuth of the sun is given by the formula:

$$\text{Cos } Z = \frac{\sin D - \sin L \sin h}{\cos L \cos h}$$

where $h$ is the true height of the center of the sun at the moment.

As a practical matter, it is not possible to collect any significant amount of solar energy with the altitude of the sun much less than 5° above the horizon. This offers some latitude in the construction of the steam collecting system: Full rotary joints are not needed because it is not really necessary to rotate the collector through a full revolution. The curve of Fig. 13-5 furnishes an approximate table of sun declination versus day of the year.

# Tracking the sun

Two fundamental approaches may be taken to track the sun. The position of the sun is well enough known so that the collector can be precisely calibrated and the

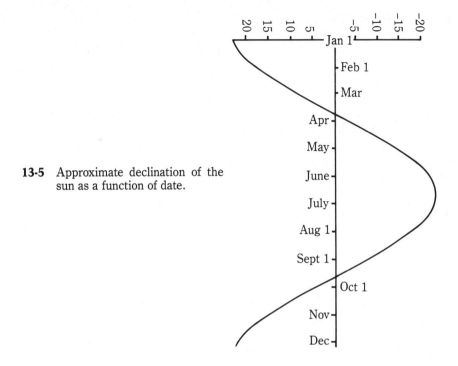

**13-5** Approximate declination of the sun as a function of date.

reflector driven to track the sun with no sensing. However, this implies a knowledge of the direction in which the collector is pointing to a very high level of accuracy. The pointing angle is affected by flexure of the mount and a variety of other errors. In order for the preset tracking arrangement to be sufficiently accurate, it would be necessary to compensate for each of these.

A far more practical mechanism for tracking the sun is to employ active sensing. At any time that the sun is visible enough to provide any significant amount of solar energy, it is a relatively simple matter to sense the pointing angle of the collector and drive the unit to an optimum pointing angle.

Figure 13-6 illustrates one such technique. The shadow of the boiler is allowed to fall on a pair of solar cells so that when the collector is properly pointed, each

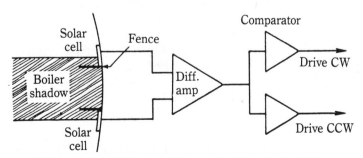

**13-6** A simplified servo drive system for solar tracking.

cell is about half covered. The output of the two cells is then compared and used to derive a signal to drive the collector in one or the other direction to restore the signal balance for the two cells. By the use of a pair of suitable baffles or fences, the servo can be made to function even when the sun is so far off aim that the shadow of the feed misses both cells. The design of a servo of this type is somewhat beyond the scope of this text.

It should be noted that the position of the shadow of the boiler upon the reflector is a very sensitive measure of tracking. Therefore, the system can be made to track with very suitable accuracy. Because the servo is actually sensing the final parameter to be optimized, the effects of errors and flexure in the mount and the collector itself are essentially eliminated.

# System stability

The simplified servo drive of Fig. 13-6 is, of course, far too simple to function effectively in a continuously operating solar tracking system. In the first place, the system must have some significant amount of deadband in order to prevent the system from continuously hunting or jittering. This means that there must be some small angle where the servo is completely shut off and does not attempt to drive the reflector in either direction.

As has been noted earlier, the reflector and altazimuth mount assembly will be a rather large and heavy structure. The motor and drivetrain used to drive the assembly to track the sun will generally have a very large gear reduction. For example, suppose that the wheels have a *rolling radius* of one foot (rolling radius is the effective radius that describes the forward motion of the wheel under rated load and at rated tire pressure, and it may be found by measuring the forward progress of the wheel in one complete revolution and dividing by $2\pi$). If we assume that the radius from the azimuth axis out to the center of the wheels is 12 feet, then the circumference of the azimuth circle is $2\pi \times 12$ feet or 75.4 feet. With a rolling radius of one foot, the wheels will advance 6.28 feet per revolution or will require 12 revolutions to complete 1 circuit of the azimuth circle. Stated differently, the azimuth angle of the reflector will change by 30° with each revolution of the wheels. At noon in midlatitudes, the required azimuth rate for the reflector is about 15° per hour. Therefore, the wheels will be turning at the rather slow rate of a half revolution per hour.

Let us suppose that we wish to drive the reflector in azimuth with an ordinary induction motor that turns at 1750 rpm. We would probably like to have a capacitor-run motor because of the ease with which reverse operation can be obtained. If we choose to have the motor capable of slewing the collector assembly at a rate of 45 degrees per hour, our average motor speed would be about 583 rpm, and we would require a gear reduction ratio of about $583 \times 60$ min per hr $\times 2$ hr per rev = 70,000:1 between the motor shaft and the wheels.

There are several facets to the use of such a large gear reduction. First and foremost is that it will not be necessary to use a very large motor to drive the collector because the huge gear reduction will multiply the torque of even a small motor up to levels adequate to move the reflector, provided that it will roll freely.

A second and less desirable facet of the very large reduction comes from the multiplication of any play or looseness in the gear train. If we presume that a looseness or "slop" in the gear reduction is to be held down to a total swing in aiming of the reflector of a quarter of a degree, this would be multiplied by 12 to obtain the looseness in a given wheel and again by 70,000 to obtain the looseness measured at the motor end of the train. This amounts to 210,000 degrees of rotation of the motor or 583 revolutions. If the motor were driving the unit at standard rate in one direction, and the unit stopped and the motor then reversed to cause the unit to back up, the motor would run for 20 seconds in the wrong direction before it had taken the slop out of gear train and the collector began to move in the opposite direction!

It is possible to obtain a 70,000:1 reduction using a small 100:1 worm gear speed reducer driving a second unit with 100 times the torque rating and a second 100:1 reduction and perhaps obtain the final 7:1 reduction using a timing belt. But this arrangement would not prove too satisfactory in practice for a solar tracker since worm drive reduction units tend to have a very large lost motion or play or slop. A far superior arrangement would employ one of the harmonic drive units that can be obtained commercially with reductions in excess of 1000:1. These units tend to have inherently less lost motion than a worm drive. An equivalent advantage exists for the use of a timing gear drive compared to a chain at the wheel end of the reduction. The timing belt arrangement can be made with relaxed tolerances and a take-up pulley employed to take out the majority of the lost motion. For truly satisfactory solar tracking it is probably necessary that the drive reduction gear train be tight enough so that the total lost motion at the motor end represent no more than about five seconds' running time at the motor end at operating speed.

In a practical system of this sort, it is not usually a good idea to attempt to achieve proportional control over the motor. The simplest and most effective servos for a drive of this sort usually turn the motor completely on or completely off. The motion of the reflector will usually represent a series of very small jogs. If the operation of the system is otherwise smooth, the motor will cycle on and off without the motor's ever coming to a complete stop.

# Play and sway

While it is necessary to have the drive capable of slewing at a rate some multiple of the maximum tracking rate, it is not advisable to have the overall system too "zippy." In the normal operation of a tracker such as the solar tracker, the target will frequently be lost behind a cloud or otherwise obscured. During this period tracking errors will have the opportunity to build up. When the sun reappears, it is necessary that the collector be able to go significantly faster than the standard tracking rate to give it a chance to catch up. This catch-up rate must be so much faster than the normal tracking rate that the image of the sun or a fraction of the image will not remain focused on the noncooled portions of the feed structure. At the solar concentrations required for a good solar boiler design, the heat of the image of the sun focused on the boiler support struts could be adequate to cause severe overheating, distortion, and in the extreme case, melting of the structure. The boiler is equipped to take the heat away whereas the support structure is not.

On the other hand, a system capable of providing accelerations that are too rapid is not desirable since it can cause severe mechanical stress in the collector and support. The boiler, reflector, or refractor frame and the elevation mount will all have a certain amount of elasticity. This can give rise to system instabilities in several ways.

For example, consider the arrangement shown in Fig. 13-6. Suppose that the unit lost the sun behind a cloud for a period of several minutes and that thereafter the sun emerged and the top cell was completely exposed and the lower cell completely shadowed. The servo would immediately begin driving the boiler upward to equalize the difference, probably at top speed. When the shadow was again equalized, the motor power would be removed and the motion would tend to slow down. However, the system has some substantial amount of inertia, and there must always be a certain amount of flexibility in the azimuth suspension, the support structure, the collector itself, and the feed structure or boiler support. Even if the motor were to stop instantaneously, the flexibility in these components would permit the entire assembly to swing somewhat past the mark.

For example, let us consider a unit with a 10-foot focal length, where the sensing solar cells are 2 inches in diameter. If the total overswing at the feed were as much as 2 inches (0.95°), the upper cell would be completely shadowed (briefly), and the drive motor would be turned on in the reverse direction. At just about the same time that the collector was trying to swing back toward the neutral position, the drive motor would have taken the slack out of the system and would be driving the collector full speed in the downward direction. This drive could cause the feed to whip to the low position again, where the motor would be excited to drive the entire system upward and the system could go into a series of oscillations, which might easily build up to the point where some portion of the structure would mechanically fail.

The calculation of the actual spring constant and moment of inertia of a structure as complex and large as a working solar collector is an enormously complex problem, which would tax the resources of a fair sized mechanical engineering group and is probably completely out of the reach of most do-it-yourselfers. For this reason, an a priori prediction of the criteria for stability of a given system is probably not feasible. However, this does not mean that a home-brew system cannot be made to work. It simply means that it will probably be necessary to determine experimentally some of the parameters of the system, after the construction of the collector. Certain rules of thumb can also be applied to determine whether the system will be stable.

As an example, let us turn back to our 10-foot focal length system. If we presume that the system pointing accuracy must be maintained within 0.25° and that the effective center of swing of the system was at the vertex of the reflector (or at the surface of the boiler for a refracting system), we may perform a few simple calculations to determine the required stiffness of the system. At a radius of 10 feet, an angle of 0.25° describes an arc of 0.52 inch. If we arrange the deadband for our comparator so that the motor cuts off while there is still a 0.52-inch gap between the center of the feed shadow and the axis of the system, we will have arranged the system to have the maximum amount of deadband gap consistent with the tracking accuracy required. The system will shut down and coast as soon as possible (for a simple system).

There is a rule of thumb that can be applied to simple on-off or "bang-bang" servo systems. It says that the system will generally be stable if it coasts to a velocity less than half of the initial velocity when it hits the far side of the deadband. If we assume that the swaying motion of the entire collector system is sinusoidal with time, then this can be shown to correspond to an overswing of no more than 15 percent of the deadband offset. Stated differently, if the motor shuts off when the shadow reaches the point 0.52 inch *before* the axis, and if the overswing does not permit the shadow to travel more than 0.52 × 1.15 inches *beyond* the axis, the system will probably be stable.

It should be noted that this rule applies only if the motor responds instantaneously. Since we have seen that this is apt not to be the case because of lost motion in the drive, the actual tolerance must be somewhat smaller.

One of the assumptions in the previous discussion should be considered further, namely the center of radius of the overshoot swing. Figure 13-7 contains the assumption that the entire collector mount and assembly are perfectly rigid and that all of the overswing is due to the flexing of the tires in the azimuth mount. Here it can be seen that the effective center of the overshoot is very near the center of the azimuth mount. The overshoot positions are shown dotted. In actual practice, there will be a number of sources of flexure in the system and the actual travel

**13-7**  The effective center of overswing. The assumption is made that the entire collector system is perfectly rigid and all overswing comes from the flexure of the tires in the azimuth mount.

of the collector might not be in an arc of a circle. When conducting overshoot tests, the actual travel of the image is the most significant item and should be used as the criterion for the test.

Obviously, the distance that the overshoot tends to cover is a function of the speed at which the reflector was slewing. The slower the reflector was traveling when it entered the deadband, the less likely it is to coast too far. The restriction on the maximum slewing speed is therefore an effective way to contribute to the stability of the overall system. If, in the course of initial tests of the system, the overswing is found to be excessive, either the system must be stiffened or the slew rate reduced.

It should be noted at this point that inertia effects are not the only ones that can cause the system to overswing. The wind can also contribute to this problem. In initial tests, if the wind-induced sway is found to exceed more than about half of the pointing tolerance of $0.25°$, then either the system will have to be stiffened or the operation of the system on windy days should be restricted.

# Variable deadband systems

One obvious way to reduce the tendency of the system to oscillate or hunt is to make the deadband a function of the speed of the system. If the actual attained speed of the system is measured and used to adjust the deadband, the restriction on the slewing speed can be raised somewhat. The unit will then shut down the motor somewhat earlier and allow it to coast somewhat further to come to a stop. After the collector has slowed down and settled, the narrow deadband required by the tracking accuracy requirements can be restored and the collector will be slowly jogged into place by the drive. This is probably the most cost-effective way of achieving satisfactory performance from the overall system, since the additional complexity in the electronics is far less costly than the structural work required to provide the stiffness in the overall collector system.

The most cost-effective compromise will probably find the stiffness of the overall system determined by the wind sway requirements and then have the remaining sway allowed for by the increase in the sophistication of the servo system. The design of the velocity-sensing servo system is beyond the scope of this text. The reader is referred to a text on motor controls or other references on the subject.

# 14

# Solar collector construction

AT THE TIME OF THIS WRITING THERE ARE A FEW FIRMS ATTEMPTING TO get into the business of constructing solar collectors suitable for steam production. However, standard units with published prices are not yet in production in the 20- to 35-foot diameters required for production of the amounts of electricity that will fully satisfy the needs of the typical American home. It is possible to obtain quotations for units, but these generally are furnished only when a full set of specifications is submitted and a genuine interest in buying is evidenced.

If we consider satellite terminals of the same collector size equipped with mounting and drive, we tend to find that a 30-foot diameter reflector will run in the price range of $50,000 to $150,000. It should be noted that such reflectors are not directly suitable for solar collection since they generally do not have an optically mirrored surface. In general, the tolerances on the surface contour are in the same region as required by the faceted reflector described in chapter 12. The rework of a satellite terminal or radar antenna would be cost-effective only if such units were obtainable as surplus at distress prices. For these reasons, it seems likely that, at least for some time to come, it will be necessary for the home electrifier to build his own solar collector.

Compared to a high-speed windmill of large size or a significant water impoundment, a solar collector presents few dangers to persons not in the immediate proximity of the unit. The dangers of boiler explosion have been noted earlier and the dangers of steam leakage will be discussed subsequently. However, for the collector itself, the principal danger is that the unit might blow down or collapse under the weight of a snow load. Neither of these dangers extends very far from the immediate vicinity of the collector. For this reason, the design and fabrication of a solar collector is probably not outside the capability of the amateur designer and builder, provided he lives in an area where failure of the unit will yield damage confined to his own property.

In many respects, the construction of a solar collector has requirements similar to the construction of a small aircraft. The collector must be relatively precise and regular and exceedingly stiff for a structure of its size and weight. It is furthermore usually desirable to make the structure as light as possible. The collector structure must be supported by the elevation gimbals and be moved by the elevation drive. The elevation axis fork is then placed upon the azimuth motion assembly, which must rigidly support and propel both units. An extra ounce in the collector proper will add a half pound to the weight of the fork, which will add three pounds to the azimuth drive! Therefore, it is not unusual to find that the construction of large collectors will follow aircraft techniques.

# Reflector construction

The typical large reflector is generally constructed with a back frame to give the required rigidity and a precision front surface that is determined by a mold of some

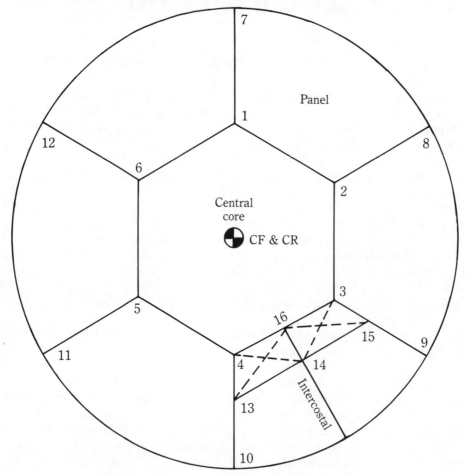

**14-1**   Subdivision of the reflector along the focal axis.

sort. The backframe is often constructed from aluminum or steel tubing in an elabo-rate truss work to provide the features of rigidity combined with lightness. The basic attempt is to divide the structure into a series of cells fabricated from triangles.

Figure 14-1 shows the view of a reflector of this construction from along the focal axis. It may be seen that this reflector has been divided into a central hexago-nal core and a series of six panels. For economy of manufacture, the panels are identical. The radius of the central core is 40 percent of the radius of the reflector itself. As the ratio of focal length to diameter grows larger than the value of 0.3 shown in the example, it is usually wiser to reduce the ratio of central core to over-all diameter.

Figure 14-2 shows the structure of the backframe in diagrammatic form. This view of it is shown along a plane going through the vertex and one of the panel dividing points. Points A and B represent the point of attachment of the panel to the central core. If the panel is loaded by a wind blowing in the direction from the focus to the vertex, it can be seen that stringer LKHEB is in compression and stringer LJFDA is in tension. Since the stringers are relatively slender with respect to their length, it is necessary that something be done to keep the stringer that is in compression from buckling. The spacers DE, FH, and JK serve to hold the stringers apart, and the diagonals AE, EF, and FK serve to turn the trapezoidal areas made up of the stringers and the spacers rigid. Since a wind-loading from the rear of the reflector is equally likely, it is necessary that both stringers be capable of operating in compression.

The advantage of the panel construction is the fact that the mold can be made much smaller than would be required if the reflector were made in one piece, and much less indoor assembly space is required. In addition to this, there is the advan-

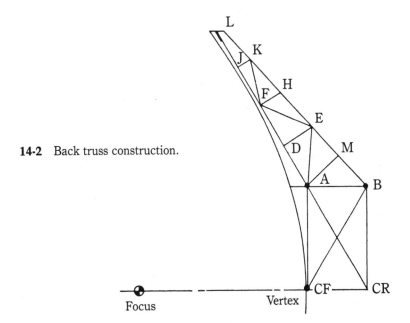

**14-2** Back truss construction.

tage that two men can handle the sections, whereas the finished reflector will certainly require the services of a crane, and a rather large one at that.

On a structure such as this, the builder has a variety of options for the truss construction. The trusses could be made of thin-wall steel tubing with the joints welded or brazed or somewhat thicker wall aluminum tubing with the joints heliarc or TIG welded or the structure can be made of wood with glue, gussets, light nails, and wire bracing. The most popular choice among home aircraft builders is the wood- and wire-braced structure since this will usually yield the highest ratio of strength and stiffness to weight using techniques available in the well-equipped home workshop.

For a wire-braced wooden structure, it is fairly common to have a ratio of truss depth to diameter on the order of 25 percent. Operating on these proportions, it is instructive to make a few calculations regarding the size of the stringers required. We will assume that the construction material is aircraft grade spruce, which has a working stress in compression of about 768 lb per in$^2$.

The panel has a total frontal area of 116.3 feet for a 32-foot diameter dish, and the center of pressure is 5.51 feet above the plane described by AB. Assuming that the wind's "fastest mile" in the area is 90 mph, we shall consider a 100-mph design. At this velocity, the wind pressure, considering the panel to be a flat plate, amounts to 42 lb per ft$^2$. This gives us a total force of 2442 lb and an overturning moment of 3364 lb at each rib.

The force vector diagram of Fig. 14-3 shows us the forces acting upon the attachment bolt at B. The shear force upon the bolt is 4600 lb, which is equal to the compressive force in stringer LKHEMB. For this compressive force at 768 lb per in$^2$, we would require about 6 in$^2$ for the stringer, which would probably be in the form of a 2 x 3 (finished) member. If we take the working shear stress of an automotive quality bolt to be 10,000 psi we find that the bolt will require a cross section of about 0.52 inch. We would probably use a $^9/_{16}$-bolt (this assumes a double-shear clevis arrangement).

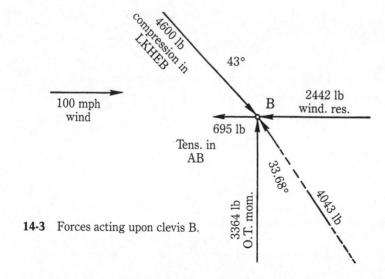

**14-3**  Forces acting upon clevis B.

You will note that in this analysis, it was assumed that the entire resistance to the wind force that would act to slide the panel straight back is furnished by the downwind clevis. Furthermore, this implies a moderate amount of tension in the member AB. If the actual tolerances in the fit of the bolts and the clevises has been such that the upwind clevis can share some of the wind resistance load, the tension will have risen in member AB. However, the compressive stress in the stringer LKHEMB will have remained constant.

Going now to clevis A, we observe in the force vector diagram of Fig. 14-4 that both the total shear forces on the clevis bolt and the tensile and compressive stresses in the members intersecting at A are somewhat smaller than those at B. If A had shared somewhat more in the assumption of the sliding component of wind resistance, the tension in stringer LJFDA would have risen and the components in members AE and AM would have fallen.

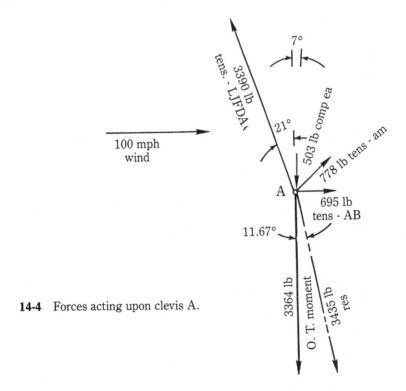

**14-4**  Forces acting upon clevis A.

Next, let us consider the case where the wind is blowing from the rear of the reflector with the same 100-mph velocity. The vector diagram at clevis B again assumes that clevis B will assume all of the sliding load. This vector diagram is shown in Fig. 14-5. It will be seen that all the vectors are reversed and that member AB now carries a compressive load of 695 lb. The loading in stringer LKHEMB is now reversed into tension, but the magnitude is unchanged.

The situation at clevis A (Fig. 14-6) is also very slightly changed except for the reversal of tension and compression. From an examination of this data it would

4600 lb tens. in LKHEMB

43°←

100 mph wind

B

695 lb comp. in AB

2242 lb wind res.

33.68°

3364 lb O.T. moment

**14-5**  Forces acting upon clevis B with the wind from the back of the panel.

seem likely that the forces in stringer LJFDA will not rise as high as those in the rear stringer and that the cross section of this stringer might be reduced to some value like 4.7 in² cross section or perhaps 2×2.5 inches. Each of the braces and diagonals is subject to considerably smaller loading and could therefore also be reduced. Also the bolt at clevis A could be smaller than that at clevis B. However, the weight savings here would be negligible.

It is fairly obvious that the rib structure would be fairly stiff and rigid. Yet we have postulated a structure in which the entire loading from the panel is carried by four clevises at the bottom edges of the panel. The actual load is distributed across the panel and there would be a tendency for the panel to sag in the center if the panel were not infinitely stiff, in the direction of the circumferential members. The panel would also be subject to racking stresses in certain positions. Referring back to Fig. 14-1, we see that panel 3, 4, 10, 9 has been shown with an intercostal rib, which is between the mounting points at vertices 3 and 4 of the central core. The crisscross dotted lines represent wire braces added to stiffen the panel and to prevent racking.

Figure 14-7 shows the effect of the wire bracing to hold up the intercostal between points 13, 14, and 15 at section EF of the rib. It is presumed that the entire bending moment load is born by the tension in the wire, the compression in

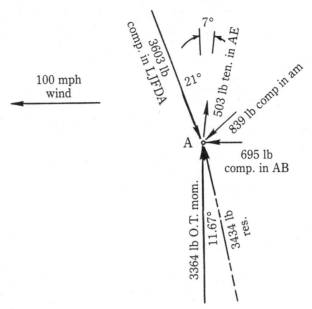

**14-6** Forces acting upon clevis A with the wind from the back of the panel.

the intercostal spacer, and the compression in the member E13, 14, 15. The load on each of these members is relatively modest. If we presume that the wire is music wire with a working tensile strength of 30.000 lb per in², then the cross section of the wire will be 0.128 inch in diameter.

This type of construction will yield a very rigid structure with minimal weight. Calculating the weights of the stringers and the spacers with due allowance for gussets, glue, etc. produces an estimated weight for a single rib of about 60 lb. With the total weight of the two ribs, the intercostals, and the wire bracing and hardware, the panel framework should be about 250 lb.

The design presented in this discussion is by no means the most efficient possible. If the intercostal were also attached with clevises to the central core and allowed to bear some portion of the wind resistance and the overturning moment, all three of the ribs could be lightened and the net weight of the panel would fall. As a matter of fact, in the limit, the lightest structure would have, in effect, an infinite number of ribs and attachment clevises. In this case, the design would become a *monocoque design* and would be about the lightest achievable with the strength.

In actual engineering practice, skin/foam/skin or honeycomb structures do come very close to the theoretical limit predictable for a true monocoque structure. For molded fiberglass structures with a honeycomb or foam filler, the strength to weight ratios are very high. Unfortunately, these structures are not very easy for the home builder to produce, and they generally require a substantial amount of tooling and no little amount of fabrication experience. It is expected that if commercial demand were to arise for solar collectors, these techniques would be very commonly employed in their construction.

14-7   Wire bracing of the intercostal.

# Fabricating the mold

For a shaped body such as a parabolic reflector and particularly for a paraboloid of revolution, it is nearly imperative that a mold be available for assembly and surface molding and so on. The mold must be built to the tolerances required of the final produce since the mold tends to establish the actual tolerances of whatever is built. The question therefore arises how the home builder can make a mold with the required precision for a complex curved shape without a number of gigantic tools.

One technique that has been successfully employed involves building the mold on the ground with the focal axis vertical. The mold must generally be built indoors for protection from the weather. This implies the presence of a rather large barn or garage or some similar building.

The first requirement for this structure is to have a reasonably smooth and flat concrete floor that is thick enough to take the bolts and other attachments. The floor must also be strong enough to take the loads that will be placed upon it without sagging, settling, or cracking. Before the project is finished you will have 20,000 to 30,000 lb loaded upon it. The structure should also be placed so that the mold floor will remain dry during both summer and winter.

It is also handy if the structure can be arranged so that an overhead block and tackle arrangement can be rigged since you will have to be carefully lifting and positioning articles that will weigh upward of 500 lb. A triple fall block and tackle using 3/4-inch rope will permit a fairly strong man to lift and position items up to about 1000 lb single-handedly. A second block and tackle is also handy since a double pick on the workpiece will permit level lifting even when the exact center of gravity is not known.

# The center pivot

The first step in the construction is to bore a hole through the floor and place the center pivot. This center pivot is a pipe, upon which we shall swing the cutting apparatus that will shape the mold. The center pivot should be sturdy enough to bear a considerable load. The pivot could be made from ordinary plumbing pipe with the upper section turned in a lathe to assure roundness and smoothness or it could be thick-wall steel tubing stock purchased from a local steel supplier. Our mold will eventually be a keyhole-shaped structure with the round part slightly larger than the central core and the slot portion slightly larger than the panel. The pivot should be located so that you can walk freely around all sides of the mold.

Figure 14-8 shows the general layout of the structure that will be eventually constructed. In the layout of the workspace, it should be remembered that it will be necessary to machine the core area on both sides of center. Therefore, the building must have a floor area large enough to permit the mold sweep to be walked in a full circle about the pivot.

In the first step, the socket for the pivot should be accurately located and cemented in with the pivot as precisely vertical as possible. The pivot itself should have no more than 0.015 inch of bow along the length, and the pivot should be vertical to within 0.015 inch. In order to assure these tolerances, it might be necessary to guy the pivot at the top with a series of cables so that the requisite vertical position can be adjusted and maintained. In this matter, the use of four guys rather than three will make for much easier leveling. Two people on opposite sides of the mold can slack off and take up on the turnbuckles to level the plane of the guys. The two people then walk 90° around the mold and level the guys in the other plane. A little interaction will exist between the sets, and only a few repeats to "touch-up" will suffice to make the pivot precisely plumb.

The next step in the construction entails the accumulation of a substantial amount of good, dry hardwood. Nearly any wood could be used, but maple is probably the best choice in terms of hardness and stability. It is most important that

**14-8** The mold.

this wood be aged and dry and that a single type of wood—preferably from the same tree—be used in the construction. The wood can be in random lengths and does not need to be furniture grade, but segments with knots and twisted grain are to be avoided. In rural areas, it is not too difficult to find a farm sawmill where the local farmers will cut up trees for sale as construction grade lumber. The wood should have air-dried for at least a year and a half and it should have spent the last six months in an environment with humidity similar to that of your shelter. If it has not, you would be well advised to purchase the lumber and allow it to dry, stacked with spacers between layers, within your building. This is intended to assure that the mold will not warp or split after you have spent time and effort to machine it to the reflector contour. A six-month drying period is not a bad idea in any event and you can use the time to work on other facets of your solar electric power system.

Once the wood is duly seasoned, a good plan is to rent or otherwise obtain a large thickness planer and plane all the blocks to the same thickness. The exact thickness is not too important. If you study your supply of wood you will find that there is probably some optimum thickness that will maximize the yield. Thicknesses less than 1.75 inches or greater than 3 inches are best avoided. If the wood is too thin you will have too much gluing to do. If the wood is too thick, you will have trouble with voids and gaps.

In the course of the thickness planing, the convex side of a bow should be removed first to obtain maximum straightness. The wood should be planed on all four sides and the ends trimmed off square with a radial saw.

The next step in the procedure is to lay down a sheet of 6-mil polyethylene on the cement to seal the moisture. This step is important and should *not* be neglected. The wooden billets are next laid upon the floor and the construction of the mold blank begun. Alternate layers of billets are laid cross-grain to one another after the area is painted with a good grade of casein glue. It is a good idea to test the casein glue with your water. With some types of softened water, certain casein glues will not form a brown, sticky syrup but rather a gritty, malted-milk color mix with very little holding power. Before starting the layup it is advisable to test-glue a few samples to make sure that your glue will work with your water. If problems are experienced, you can easily purchase distilled water for glue mixing.

The mold is simply laid up like an igloo with the layers and adjacent billets glued together. At the end of any day's work, a sheet of polyethylene can be thrown over the pile, and the pile can be weighted down with 50- or 100-lb sandbags. It will probably be a good idea to have a dozen or a dozen and a half sandbags around for this purpose. The best type of sandbags are the military cloth type rather than the commercial paper bags in which sand is usually sold. The bags should not be tightly filled so they are stiff. Instead, they should be limp so that they will conform to the shape of the pile.

Eventually your igloo will begin to take the shape of the mold. Precision is not particularly important at this point. You should only strive to get the shape and size of the pile such that you can machine the mold out of it. A little care in laying up the pile initially can save some machining time later. Once the pile is complete, it is advisable to give the entire thing a month or so to cure before starting any machining on it.

Once the pile is in place (and definitely not before) you can begin laying in the level track. This track is most easily constructed by using precision ground steel bar stock about $2 \times 3/8$ inches. This stock can be purchased from a steel supply house in lengths of three feet. The circular track will be approximated by a polygon with sides equal to the bar lengths. Locations for the track mounts can be laid out by a swing wire from the pivot post.

Figure 14-9 shows a fairly simple track clamp arrangement. A short length of channel is brazed or welded to the top of a length of angle iron. The latter piece is drilled to accept two bolts sunk into the floor and held in place with a casting of sulfur or lead poured into the holes drilled in the concrete. These will hold about equally well. The sulfur is cheaper and requires less heat to melt, but it is also more obnoxious and the fumes should not be inhaled. In either case, do not pour the molten material into the holes if there is any significant amount of moisture present; it will explode and send the molten material back out into your face. Always wear a plastic face shield while pouring.

The rail is firmly clamped between the shim and the setscrews. The shim is required in order that the rail will clamp in an upright position since the inside of the channel will always have a certain amount of draft or tilt to the inside wall.

The height of the unit can be precisely set by loosening the top nuts and running the bottom nuts up or down as required in order to level the rail. A good builder's or surveyor's level can be used to set the track sections level to within about 0.015 inch. After this process, a precision millwright's level can be used to fine level the track to about .002 inch. A bump at the joint between the rails can be avoided by having the ends of the rails overlap slightly in the polygon. The end of the mold sweep will be guided by this track. Therefore, it is important that the track be as level as possible.

Rail

Shim

Cement

Lead
or
sulfur

**14-9**   The precision track mount.

The mold sweep is shown without much detail to simplify the picture. This has to be a sturdy frame equivalent in rigidity and lightness to the ribs but larger since it must reach all the way to the center of the reflector. A trussed construction similar to the ribs is probably a good choice. This unit can be layed out on the open floor area adjacent to the mold and assembled using the techniques to be described for the ribs. The hand-cut parabolic track can be a sheet of 12 or 16 gauge aluminum. This track can be installed before cutting to the ribs.

The parabolic curve is laid out on the aluminum sheet by carefully plotting out points about every 6 inches using a surveyor's transit and a precision machinist's scale. With care, you will find it possible to place a line of center punch marks along the desired parabolic curve with a tolerance on the order of 0.005 inch ($\pm$). Figure 14-10 shows this technique.

To begin with, the anchors are driven in the concrete and the wires representing the focal axis and latus rectum are stretched out using the surveyor's transit. The squareness of the layout may then be fine adjusted using the 3, 4, 5 rule. For the latter, a distance that is some multiple of 3 is laid out on one wire and a distance that is a multiple of 4 is laid out on the other. When the wires are perfectly perpendicular, the distance between these points will be a multiple of 5. This test can actually be used to improve the precision of the layout.

In the next step, when the wires have been strung as nearly perpendicular as humanly possible, the mold sweep frame is laid upon the floor and jockeyed into position and lag-bolted to the floor. The aluminum sheet from which the parabolic track is to be cut is next jockeyed about on the frame until the vertex and the sail tip location are well up on the sheet. The sheet is then firmly screwed to the mold sweep frame. Whereas the lag-bolting of the frame to the floor is a temporary

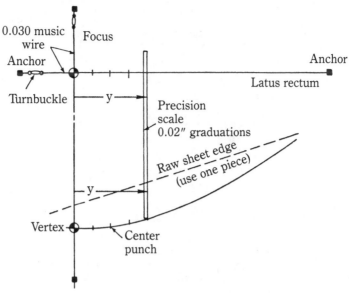

**14-10**   Laying out the parabolic track.

attachment, the attachment of the aluminum sheet to the frame is permanent, so using epoxy in the attachment is not a bad idea.

The vertex is next marked upon the sheet, and the string of points marking out the paraboloid are next carefully measured and punched with a centerpunch into the aluminum. On a large sheet such as this, it is advisable to use a blue toolmaker's ink (Prussian blue) to aid in finding and seeing the marks. When the entire string of marks has been punched in, the parabolic curve may be faired in between the marks. At this point it is not unlikely that one or more punch marks will be found that will not fair smoothly into the pattern. If this is the case, go back to the measurements and determine which points are spurious and correct them. This is especially important since any errors at this stage will be reflected in the entire final product.

Once a good true and fair curve is scribed upon the surface, the curve may be cut out using a sabre saw operating at slow speed. The saw kerf should run about 1/32 of an inch on the waste side of the curve. The curve may then be *carefully* hand filed to the line.

If you are not familiar with filing aluminum, it is a very good idea to obtain the assistance of someone who can tell you how. Get a set of new files for this task and keep the file card handy to prevent "pinning" of the files. The curve being generated will be the guide for the entire project.

The next step is the fabrication of the router cart. This unit is depicted in Fig. 14-11. The cart simply consists of a pair of rollers designed to run smoothly and with little play in a track formed of aluminum channel. The rollers are attached to a back plate and bracket designed for mounting a heavy-duty industrial router. After this has been fabricated, the attachment of the tracks can begin.

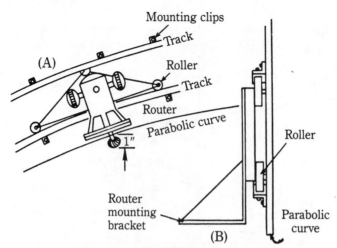

**14-11** Router cart at A and track at B.

The tracks—like the parabolic template itself—should be a single piece of aluminum (if at all possible) to avoid a noticeable discontinuity at the joints. With the parabolic curve finished, the tracks are arranged on the sheet and clamped in place in position such that the tip of a carbide ball cutter used on the router will extend a fixed distance in the X direction from the parabolic curve. The tracks may be jockeyed until this condition is obtained. The holes for the track mounting clips may then be drilled and the track attached. If a little "slop" is left in the clip holes, the track may be jockeyed a bit in the final touch-up adjustment of this assembly.

It is usually a good idea to depend upon the bottom track for the determination of the system position. There will usually have to be a little play in the tracks to prevent the cart from binding. If the bottom track is precisely located, the mold may be accurately cut despite this play by pressing down on the router.

It is important to note that the measurement to the ball cutter should be made in the X direction and *not* normal to the parabolic curve. As the router travels out from the center, it will tilt as shown in the illustration. When the track is properly installed, it will not be precisely parallel to the parabolic template because of this tilt.

With the track installed and the cart running smoothly, the next steps are in order. The first item is the attachment of a precision level to the sweep parallel to the latus rectum. The second step is the fabrication of the sweep bearings. These bearings may be made of hardwood and they should attach to the sweep so that they are offset the same distance from the parabolic curve as the center of the router. These bearings should be bored on a lathe for a smooth running fit on the pivot post. The final step is the attachment of the outboard roller on an adjustable bracket to the end of the sweep so that the roller can be made to rest upon the circular track. This roller should be cylindrical, and its axis should be as parallel to the latus rectum as possible, since an endward travel upon the circular level track is inevitable and an inclination in the axis of the roller will make for a scalloping of the reflector surface. With this installed, the mold sweep is complete. If the axis of the outboard roller and the pivot axis are both offset from the mold by an amount equal to the centerline offset of the router cutter, the tip of the cutter will describe a paraboloid of revolution as it is walked about the circular level track.

# Leveling

At this point the mold sweep may be unbolted from the floor and lifted into position above the mold. The pivot may then be inserted and the sweep locked onto the pivot with its bearings. The pivot collar is next set so that the sweep will clear the built-up block pile of the mold.

Using a precision surveyor's level, the roller is next adjusted so that the difference in height between the vertex punch mark is close to the difference in X coordinate between the most outboard punch mark on the parabolic curve. The level is then unlocked and jockeyed to read level and locked in place. For a check, the sweep may be walked about the circular track 180° and the bubble level checked. If the bubble is off, it should be reset for half of the error and the roller adjusted for the other half. The sweep is then walked back to the first position and the height difference checked. This step is important since it determines the focal length of the reflector. Any significant departure will result in a nonparabolic surface.

# Forming the mold

With the level securely locked and tracking around the entire circle, the task of cutting can begin. The collar can be lowered and the roller adjusted to give a level bubble reading with some of the block corners within reach of the router cutter. The sequence then consists of crawling about the surface of the mold pile and knocking off corners of the blocks. This part of the operation is relatively tedious and time-consuming but necessary. It is probably a good idea to have a stock of carbide ball cutters since several will be worn out before the mold is finished. Depending upon the care with which the stack was built, there might be a great deal of cutting to do in certain areas. It is probably a good idea not to try to take off more than about 1/4 inch in these initial cuts. In the initial cutting, the best tack is for the operator to climb up on the mold stack and work the router radially. As the job progresses, the mold will become smoother and smoother and it will be difficult to sit on the inclined surface without falling off. At this stage of the finishing, it is generally easier to lock the router at some fixed radius and walk the tail end of the sweep in a circle. When the cutting has proceeded to produce a smooth paraboloid of revolution, any voids that have been exposed may be filled with auto body epoxy or other filler and a final pass taken on the surface. After finishing, the surface should be sealed with shellac or other wood sealer.

# Assembling the framework

With the mold complete, the task of making up the frames can begin. The first task is that of assembling the central core. The frames for this unit can be built upon the floor and the cross framing and wire bracing installed later.

At this point the mold sweep should be removed and the pivot should be removed as well. Figure 14-2 shows the stringers and frame members straight and the curve between the parabolic surface and the framing occupied by cut formers. These cut formers attach to the ribs and frames. The outlines for these formers may be traced from the sweep and the formers sawed out of plywood with a sabre saw. The central frame is now hoisted up into position and the formers braced against the mold and glued and nailed in place to the frame. When this is finished, the clevises can be installed on the frame and the construction of the panels begun.

For the panels, it is advisable to provide a flat floor space large enough for the assembly of the intercostals. If the floor is covered with a construction plywood false floor, blocks of wood can be nailed to the plywood to make a jig to hold the pieces of the ribs in place during gluing and assembly. In most cases the intercostals will be larger and of slightly different shape than the frames. Therefore, a second set of blocks can be used for jigging these. It is usually worthwhile to make all the ribs and all the intercostals at one time since the duplication will provide certain economies of scale. All the stringers, spacers, and diagonals can be cut at a single saw setting and so on.

If the mass production of the frames is to be attempted at a single setting, it is usually worthwhile to build a dummy plywood template of each first so that the design can be checked for any gross errors before all the pieces are cut and the error is replicated throughout the structure. A single dummy for each separate rib and intercostal design will suffice.

With the pieces cut, the stringers may be laid into the jig and the braces and diagonals glued in place and the gussets glued and nailed on one side. The frame may then be turned over and the gussets glued and nailed in place on the opposite side when the glue has dried for the first. During the drying process, the members may be held in place with the sandbags prepared for the mold layup.

The basic construction of the rib joints can be seen in Fig. 14-12. The stringers, spacers, and diagonals are fitted together and glued. A plywood gusset is glued and nailed to each side of the joint. Care should be taken to use small nails in this process. The strength comes from the glue and *not* from the nails. The glue should be a good waterproof glue made for marine or aircraft use. Strongly resist the temptation to drive nails into the end grain of the spacers and stringers. In general the nails should disturb the grain of the stringers as little as possible, and the end nailing into the spacers and diagonals will only *weaken* the stringers!

Once a set of ribs and intercostals are assembled, it is possible to build a panel. To do this the formers are cut out and clamped to the ribs. The ribs are then temporarily clamped in place on the mold using the central core clevises and a falsework jig made to hold the ribs and intercostals in place. The formers are then glued to the ribs and intercostals using light nails as with the gussets. The formers should be pushed down into intimate contact with the mold surface during the gluing process. A sheet of waxed paper can be placed between the formers and the mold to prevent inadvertent adhesion to the mold due to spilled or running glue.

With the formers in place it is possible to begin adding the cross members. Figure 14-13 shows the means of attaching the cross members. An angle bracket is bent up of aluminum alloy and is attached to the cross member and the spacer with

**14-12** Stringer spacer, diagonal, and gusset assembly.

through bolts. This bracket not only serves to stiffen the joint but it provides a mounting place for the brace wire in the plane of the cross member (these are the wires running, for example, from E 13 to D 14, etc.). A second plate with turned-down ears is arranged to bolt under the corner bracket matching holes R and S. This plate is attached through the gusset with through bolts at T and U. This plate serves to anchor the diagonal wires; those running, for example, from E 13 to A 16, etc. Note that the bolt holes *do not* pass through the stringer but rather pass below it. Again, this is done to minimize the weakening of the stringer due to holes and fasteners. With these brackets in place, the cross members can be bolted in place and the brace wires taken up comfortably swag.

In order to ensure a uniform dress of the brace wires it is nearly imperative that a wire tension gage be employed, such as the one shown in Fig. 14-14. It may be purchased from an aircraft supply house or fabricated rather easily. The gage operates on the principle that the pressure of the spring deflects the wire by an amount proportional to the reciprocal of tension in the wire. The pulleys should be made fairly large in radius so that the stiffness of the wire does not contribute

**14-13** Crossbrace and wire assembly.

**14-14** The wire tension gauge.

errors. The homemade gage can be calibrated by hanging known weights on a vertically hung wire and noting the gage deflection (Fig. 14-15). For the wires in an assembly of this proportion, a preload strain of about 25 lb is adequate to ensure that the wire is not slack. Too much preload simply places unnecessary stress upon the structure.

# Installing the skin

Once the panel frames are assembled, the completed panel frame may be removed from the mold. The frame is now ready for the skin installation. There are a variety of materials that can be used for the skin, but one of the easiest to work with is expanded aluminum mesh. This is an aluminum sheet material that has been slit and stretched so that a series of diamond-shaped holes are presented between somewhat twisted square section wires. One of the chief advantages of this material is that it is stretchy.

The expanded aluminum mesh is laid on the mold over a layer of waxed paper to prevent adhesion to the mold. Because of the compound curve of the mold, it will not lie flat. This problem is cured by stretching the aluminum with sandbags hung with hooks to the periphery of the sheet until the sheet is stretched like a sweater to the contour of the mold. With the mesh stretched in place, the edges of the formers are coated with a thixotropic epoxy glue and the panel section carefully lowered back into position and left for the epoxy to dry. The tension on the mesh should be relaxed only after the curing of the epoxy is complete. The panel may then be lifted and the fastening of the skin secured by means of light aluminum wires passed through small holes in the formers and the holes in the mesh and twisted tight. This process should start at the panel periphery and proceed toward the center. When the mesh has been completely secured in this manner, the mesh can be trimmed to the outline of the panel. Care must be taken to avoid denting of the mesh.

In the event that the mesh does become dented or sags in some areas, the reflector can be placed back on the mold and the mesh can be tightened in given areas by twisting the connecting part of the mesh slightly with a pair of long-nosed pliers. This is one of the basic advantages in using the mesh. The second advantage is the fact that the openings of the mesh will give a good grip for the mirror tiles.

As a given panel is finished, it is carefully marked to match the central core. The core is then rotated to the next panel position, the panel is removed, and construction proceeds on the next panel. Before removal of the panel from the mold, it is a good idea to give the panel the required wood sealer and finish. The mesh is protected in this position.

While it is certainly possible simply to finish the wood of the reflector for outside usage and to rust-protect the wire stays, a good practice would be to cover the outside of the backframe with aircraft fabric of some sort. This serves to improve the structure in several ways. First of all, it protects the wood and stay wires from the weather and the sunshine. Unlike a sailboat, the back truss work is a rather complex structure to refinish annually and perhaps more often in very sunny climes. Secondly, the fabric enclosure will help by considerably reducing the wind resistance of the structure. This is advantageous both from the standpoint of windstorm survival and from the operational standpoint since a reduced wind resistance will make the problem of tracking the sun in high winds much less serious. There are a variety of fabrics with rated lives of 5 to 20 years in aircraft service. In general, the fabric is attached by both glue and sewing. The application of airplane

**14-15** Calibrating the tension gage.

Known weight

dope will then shrink the fabric to a smooth drumhead consistency. A standard manual on aircraft covering should be consulted before starting this job.

When all the panels have been finished, the central core can be surfaced with mesh in a similar fashion and the rear surface covered. It is probably advisable also to cover the interfaces of both the panels and the central core. In this way the structure of the backframe will be completely enclosed. This maximizes one of the advantages of the rear covering if the collector is to be used in a snowy clime in that the collector will accumulate much less snow, sleet, or ice on the smooth fabric back side than on the truss work of the frame. It is important that the units be pretty well sealed since an amazing amount of snow can blow through a relatively tiny crack or opening in a storm with dry snow and high winds.

With all the units finished, the next task is to weigh the panels and the backframe and to calculate the center of gravity of the structure. The reflector pivot and the counterweight can be made in the form of an "A" frame structure that attaches to four of the vertices of the central core, for example points 2 and 3 and 5 and 6 of Fig. 14-1. This structure can carry the azimuth axis and the counterweight that balances the reflector. If the counterweight is given a substantial flat area, equal to about one-quarter of the projected area of the collector proper, it will also serve as a wind balance for the unit in the azimuth plane and will reduce considerably the effect of wind torques upon the azimuth drive.

The collector servo drives can be made to function if the collector is not balanced about the elevation axis and is not wind-balanced about the azimuth axis. However, these balances will make the achievement of a smooth acting drive much easier to attain and will permit operation of the solar collector system in much higher winds.

# Tiling the reflector

With the mold available, the tiling of the reflector can proceed in one of several manners. The surface of the mold can be covered with the tiles laid with the reflecting surface down. The surface of the mesh can then be coated with a thixotropic epoxy cement, and the individual panels and the central core lowered onto the glass, and allowed to harden. This is a faster way to do the job than the individual tiling operation described in chapter 11 for a ready-built reflector. However, it is often not quite so accurate as a well-done tiling job where the tiles are individually set. In addition to this there is likely to be a fair amount of tile cleaning to do after the epoxy has set, unless the mix used is very non-runny. As before, the surface of the mold must be protected from adhesion.

# Joint effort

As can be seen from the foregoing discussion, the construction of the mold and the reflector construction jigs and tooling is a lengthy and time-consuming process, as well as expensive. The tooling is capable of producing a number of reflectors. Therefore, it would make sense to do the tooling as a joint shared effort and build a number of collectors using the same tooling and building structure. As described and with some care in workmanship, the solar collector should have a useful life of 15 to 20 years with a reasonable amount of maintenance. With the expanded mesh backing, the reflector would also make an excellent satellite terminal or a radio telescope at times when the sun is not shining. Therefore, hobbyists with these interests could well be attracted into a joint effort to construct the reflector tooling. During the sunshine periods in the summer, the unit will cast a very large and fine shadow suitable for picnicking. This factor might be used to soothe your family's feathers during the initial discussions of the project.

# 15

# The solar heat engine

HAVING EXAMINED THE VARIOUS FEATURES AND TECHNIQUES FOR collecting solar energy to render some working fluid very hot, we shall next take a look at the requirements of the system as a whole. A discussion of the properties of the "D-Valve" Corliss and Uniflo steam engines is beyond the scope of this text and is well treated in a number of other volumes. As a practical matter no one in the United States is regularly manufacturing steam engines in the 3- to 10-horsepower range. However, there are a few places where one can have such an engine manufactured to order for about $10,000. Several firms custom-build these engines for steam launches for hobbyists.

Another possible source of a steam engine is in the the form of kits offered for steam hobbyists. There are several magazines dedicated to the "live steam" train hobby. In particular, Coles Models Co., PO Box 788, in Ventura, CA 93001, offers kits of castings for several engines that would be suitable. Their #6A Condensing Compound has a 2.5-in. bore and 3-in. stroke. It is rated to deliver 4 brake horsepower at 750 rpm running on 120 psig steam. The machine is 20 in. high, 18 in. long, and 15 in. wide and weighs 150 lb. The set of castings is listed at $1650 in 1986. The condensing "Swan" has a 2.25 in. bore and 2 in. stroke and is rated at 3 bhp running at 800 rpm on 100 psig steam. It is 14 in. tall and 29 in. long and 16 in. wide and weighs 72 lb. This engine is available in a set of castings for $1200 (1986) or with the bed plate, bearing caps, crankshaft and balance weights, standards, steam chest disk flywheel, and top and bottom covers machined for $2500. A measure of the effort required to produce this engine is the fact that the raw castings set weighs 111 lb and the partially finished machine set weighs 98 lb. It is likely that the 13-lb difference represents cast iron chips removed! That is a lot of machining! The finished weight of 72 lb indicates that there is still a lot of cast iron to be reduced to chips!

The Swan is a two-cylinder, double-acting, single-expansion and the 6A Compound is two-cylinder, double-acting, double-expansion or compound type. This probably means that the 6A would be more efficient; that is, it would produce more power from a given amount of steam. By the term "double-acting" we mean that the steam pressure is applied to both sides of the piston; first from the top to push the piston down and then from the bottom to push the piston up. In the traditional steam engine this was usually the case.

There are a few manufacturers producing steam and gas turbines in the 3- to 10-hp range, but these units are extremely expensive with prices quoted at $10,000 and above, which essentially removes them from consideration. In addition, a typical 3-hp steam turbine is about 6 inches in diameter and rotates at speeds on the order of 100,000 rpm. It would require a very expensive and precise reduction gear drive to get this speed down to one suitable for driving a 60-Hz alternator. From a practical point of view, the steam engine to be used in a solar home electrical plant will probably consist of something converted from an outboard motor or a small car engine. There is a considerable amount of machining required to do this; however, it is small compared to machining up a steam engine kit. Of course, if one chose to do the kit he would wind up with a lovely hobby/collector's piece.

# The converted steam engine

By the time of the final flowering of the reciprocating steam engine, the engines of choice were *uniflow* types. These engines admitted the steam at the top of the cylinder and evacuated the steam at the bottom of the stroke through a series of ports surrounding the cylinder. This action prevented the alternate heating and cooling of the top of the cylinder and provided the longest possible conduction path between the relatively hot steam inlet and the relatively cool exhaust outlet port. These actions minimized the heat losses within the engine.

It is noteworthy that the common two-stroke outboard motor is provided with a similar set of ports. The bypass from the crankcase to the cylinder, which admits the fuel, and the exhaust port are both situated at the bottom of the cylinder as is the exhaust port of the uniflow engine. One can reason that the steam inlet port (which can be much smaller than the outlet port because of the higher driving pressure) can be installed in place of the spark plug or the compression relief valve at the top of the cylinder. The principle missing item is the variable timing linkage to operate the steam inlet valve.

There are two difficulties with the conversion of an outboard motor to steam operation:

- The outboard motor is designed to be lubricated with a fuel-oil mist in the crankcase. For steam operation this mist will have to be replaced with a more positive lubrication.

- Outboard motors are designed without the oil-control rings that are used in a "wet" crankcase engine. In the outboard, it is only necessary to seal the compression since any oil on the cylinder walls will eventually be burned in any event. For steam operation some measures might have to be taken to reduce oil consumption.

Aside from these difficulties, the outboard would seem the most likely choice for a steam conversion unit. The 3600-rpm operating speed desired for a 60-Hz alternator is well within the normal operating range of these engines, and the pressures developed by the steam are much more modest than the sharp shock of the gasoline explosion.

Figure 15-1 depicts schematically the outboard-steam conversion. The piston is shown near the bottom of the stroke with the bypass and exhaust ports (which have both been connected to the condenser) open. It is of course necessary that the portion of the bypass leading to the crankcase be blocked. The upper end of the piston stroke is shown dotted. A steam valve has been inserted into what used to be the spark plug hole. You will note that the cone of the steam valve is reversed with respect to the cone used on automotive valves or the compression release valve

**15-1**  The outboard-steam conversion.

usually used on outboards, since the pressure is always higher on the steam supply side of the line than on the cylinder side. An oil-control ring has been added to the piston.

The normal reed valves have been closed off except for a small section that allows the oil mist lubrication of the engine through a capillary tube. Since the condenser will be operating well below atmospheric pressure, there will be a net leakage of air from the crankcase to the condenser and a small pulsating flow of air and oil mist at the top of each stroke will lubricate the engine and make up the loss. Both the air leakage and the oil pickup are probably unavoidable. These impurities will have to be removed from the condenser later. This is not an unusual operation, since most steam engines had air and oil scavenging provisions for the condenser.

# Speed control

As had been noted earlier in chapter 5, there are distinct advantages to operating the home electrical system at precisely 60 Hz, since many of the appliances and devices were designed for 60-Hz operation and such matters as power factor correction are dependent upon the maintenance of the frequency. Whereas a windmill might be better operated to generate dc that is subsequently converted to 60-Hz ac, the hydro plant and the steam plant are more amenable to direct drive of the 60-Hz alternator at 3600 rpm. Because this type of operation is more efficient than the direct current generation/conversion process, it seems advantageous to consider the system built in this manner. Auxiliary and standby power can be obtained from batteries charged by conventional battery chargers, or from an automotive alternator rectifier mechanically driven by the steam engine.

In the presence of a widely varying system load, the requirement arises for a fast-responding speed control mechanism for the steam engine. With a heavier load the engine will tend to slow down and the output frequency will fall unless a control system is available to rapidly increase the output of the machine.

Traditionally, the speed of steam engines has been controlled in power applications by means of a throttle valve and by the adjustment of the valve linkage. The throttle valve was usually controlled by the governor, and the valve linkage was generally a manually operated affair adjusted by the engineer to account for larger and slower variations in the load. On steam turbines, the throttle valve is essentially the only mechanism for speed control.

For the small home electrical system, this type of operation is somewhat less desirable than it is for the large power system. For one thing, the home electrical system will experience much larger changes in demand as a percentage of installed kVA than will the large system. Furthermore, the small system will generally not have the benefit of the pull-in capability of a large number of synchronized machines. The small home system must therefore be capable of sustaining a much faster response rate to changing load.

# Electronic steam injection

To meet these requirements with maximum efficiency, the writer is prone to favor the use of a very modern technique analogous to the electronic controlled fuel injection that has begun to make its appearance on sophisticated internal combustion engines. In Fig. 11-7, a rather idealized pressure versus volume curve for a vapor cycle engine was presented, and it was noted that the area under the curve represented the work done by the engine. It should be relatively obvious that lengthening the segment AB in this curve will increase the area of the curve. It also represents the consumption of more steam. The length of segment AB represents a measure of the length of time that the inlet valve was admitting steam to the cylinder. More steam equates to more work output. Similarly, shortening the curve segment AB would represent admission of less steam to the engine and therefore less work. An electronically controlled solenoid valve for steam admission could be arranged to permit control of the steam admission on a stroke-by-stroke basis and would thus provide the fastest response possible to changing load requirements for any given engine.

# Inlet valve timing

The older steam engines made use of a rather large ratio of connecting rod length to stroke or crank throw. If the connecting rod were infinitely long, the motion of the piston would be simple harmonic motion; that is, the graph of piston position versus time would be a sine wave. This has several advantages:

- The vibration moments would be confined to a single frequency equal to the crank rate.
- The piston would spend an equal time at each end of the cylinder. This was important in a double-acting engine in which both ends of the cylinder were used.

Modern internal combustion engines have gone in the direction of the use of a very short connecting rod compared to crank throw since this tends to make the engine much more compact. Furthermore, no modern high-performance engine makes use of a double-acting design. The harmonics introduced by the non-simple-harmonic motion of the piston have been canceled by the introduction of sophisticated balancing techniques.

A difference is brought about in the "indicator diagram" or P versus V diagram of the engine by this design change. Figure 15-2 shows an indicator card diagram more typical of the results to be obtained for an outboard-steam conversion. The horizontal axis has been plotted to show two scales. One of these is the percentage of full cylinder volume and the lower one is the angle of crank rotation. From the second curve it may be ascertained that the engine spends a good deal of its time near minimum volume and even more of its time near maximum volume with the period in between being traversed very rapidly. The rotation curve was calculated assuming that the engine has a compression ratio of 6:1 and that the connecting rod length is 2.5 times the crank throw radius. These figures are representative of modern high-performance internal combustion engine design.

The figure shows one curve in full representing the nominal load indicator card. Also shown dotted are two curves representing low- and high-load operation.

**15-2**   Outboard-steam conversion indicator cards.

In the low-load condition, the inlet valve was open for such a short period that the pressure in the cylinder did not have time to approach the boiler pressure and the valve is performing a time-modulated throttling action. At the nominal-load condition, the valve was open long enough for the cylinder pressure nearly to reach boiler pressure. In the high-load condition, the valve was held open long enough for the engine to hold near-boiler pressure in the cylinder for about 35° of crank rotation after top dead center.

If we assume that the engine is turning at 3600 rpm, then the crank is making a full revolution every 0.0167 seconds (or 16.7 milliseconds) in the low-load condition, if we wanted to have the inlet valve open 10° before top dead center (TDC) and reclose about 15° after top dead center. The valve would be open for 1 millisecond. At nominal load, if we wish to open the valve 15° before TDC and reclose it 25° after TDC, the total period would be 1.85 milliseconds.

This is a very fast action indeed, comparable to the opening time of a high-speed camera shutter. However, similar performance has been attained in cold-gas thruster valves employed in the maneuvering mechanism for spacecraft.

A valve to operate at such speed can be constructed with an ordinary poppet-like mechanism driven by a moving coil solenoid of the type used in loudspeakers, but this would require a considerable amount of electrical energy. Figure 15-3 shows a servo valve that can be operated with considerably less electrical drive.

**15-3**  The servo inlet valve.

In the servo valve, a steam inlet enters a cylinder containing a spool. The body of the spool is shown blocking the outlet, held in place by a spring. Since the steam pushes equally on both ends of the spool, the spool is in equilibrium. If the small solenoid marked *solenoid 1* is briefly pulsed, the needle valve releases the pressure in the bottom chamber and the valve spool will snap down to a new equilibrium position with the steam outlet open. Solenoid 1 may then be reclosed. When the time arrives to reclose the inlet valve, solenoid 2 is briefly opened and the pressure in the upper chamber falls. The restored pressure in the lower chamber overbalances the reduced pressure in the upper chamber and flings the spool up again. In effect, the valve itself is a small free-piston steam engine. If the pressure releases are placed somewhere up from the end of the piston, as the spool passes the release outlet, the remaining steam is trapped in the end chambers. The compression of this steam prevents the spool from slamming violently into the end walls of the chamber. Naturally, the spool must be precisely lapped to fit the valve cylinder in order for the valve to operate without much leakage.

Another item is worthy of attention. An ordinary auto or outboard engine has the compression ratio limited in order to prevent detonation or knocking in operation. No such limitation exists in the steam engine. The *clearance volume* at the top of the cylinder simply absorbs steam. Since the amount of steam in the cylinder is larger, this would tend to make the diagram fatter. A closing off of some of this clearance volume by the addition of a block to the cylinder head will improve the operation of the engine with respect to the maximum-to-minimum output ratio and will improve the efficiency of the engine, particularly under light loads. The reduction of clearance volume is best accomplished on the cylinder head, where it does not affect the operation of moving parts and upset the balance and vibration damping of the engine.

# The solar power system

The complete solar power system is depicted schematically in Fig. 15-4. We shall briefly discuss how each of the items is used in the system and its requirements.

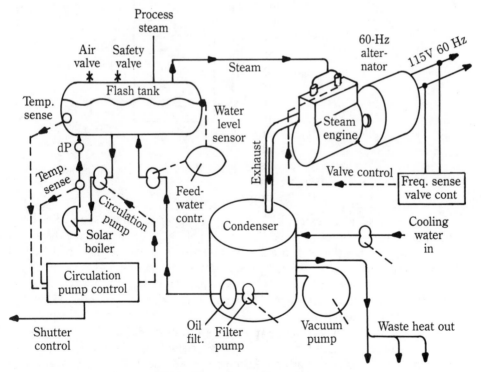

**15-4**  The solar power system.

# The circulation pump and control

As noted earlier, the solar boiler was designed to contain a minimal volume of water. The reason for this is to permit the boiler to respond by providing usable heat energy in very brief periods. When the sun has ducked behind a cloud and the boiler cools, the circulation pump is shut off. Conversely, as soon as the boiler starts to produce water temperatures that would contribute to the flash tank, the circulation pump is operated to keep the boiler from overheating. The little item marked dP is a pressure constriction intended to ensure that the pressure within the boiler is always slightly higher than the pressure within the flash tank. This prevents the actual boiling of the water within the boiler proper. This addition is necessary because the boiler will probably be installed at some level higher than the flash tank.

In more common steam boiler designs, the difference in elevation between the higher flash tank and the boiler tubes is used to provide this differential. The circulation pump will operate continuously, but at a very slow rate when the boiler is producing no heat.

# The flash tank

The flash tank serves the function of storing the hot water generated by the solar boiler and separating out the steam from the superheated water. The water enters the flash tank in the liquid state under high pressure. When the engine draws some steam from the flash tank, the pressure falls slightly, permitting some of the superheated water to flash off into steam. If this tank is made small, the system will come up into operation very rapidly. However, the system will have little stored energy to coast down, if the sun passes behind a cloud. On the other hand, if the tank is made large, it can store a considerable amount of energy. This comes at the expense of a lengthy time to get up-to-steam. If the up-to-steam time is too long, the ability to use brief intervals of sunshine is lost since the system might not get up-to-steam during the brief interval and no power might be produced at all.

For a numerical example, let us consider a system that delivers 2.24 kW of electrical energy. Let us further assume that the losses due to mechanical inefficiency of the engine, the water pumping, and the tracking of the solar reflector bring the efficiency of the unit down to 0.71. This brings the useful power requirement up to 3.15 kW or 10,768 Btu/hr. If the inlet to the engine has saturated steam at 200 psia and the outlet to the condenser is at 1 psia, we may refer to the saturated steam information in Table 15-1 and observe that the total heat difference between saturated steam at these pressures amounts to 1198 – 1105 = 93 Btu/lb. Our engine would therefore be consuming 10,768/93 = 115.8 lb of steam per hour.

# The condenser action

The condenser, which serves the function of the refrigerator in the Carnot machine, must be prepared to turn the steam back into water for repumping to the

**Table 15-1. Saturated steam.**

| Absolute pressure psia | Temperature °F | In water | Heat of vaporization | Total heat |
|---|---|---|---|---|
| | | Btu/lb above 32°F | | |
| 1 | 101.8 | 70 | 1035 | 1105 |
| 2 | 126.1 | 94 | 1022 | 1116 |
| 5 | 162.3 | 130 | 1000 | 1130 |
| 10 | 193.2 | 161 | 982 | 1143 |
| 14.7 | 212.0 | 180 | 970 | 1150 |
| 20 | 228.0 | 196 | 960 | 1156 |
| 50 | 281.0 | 250 | 923 | 1173 |
| 75 | 307.6 | 277 | 904 | 1181 |
| 100 | 327.8 | 298 | 888 | 1186 |
| 150 | 258.4 | 330 | 863 | 1193 |
| 200 | 381.8 | 355 | 843 | 1198 |
| 250 | 401.0 | 376 | 824 | 1200 |
| 300 | 417.3 | 394 | 808 | 1202 |
| 400 | 444.6 | 424 | 780 | 1204 |
| 500 | 467.0 | 450 | 754 | 1204 |

boiler. In this case, the table tells us that we must remove 1035 Btu/lb from the steam at 101.8°F in order to condense it into water. This amounts to 115.8 lb/hr × 1035 Btu/lb = 119,832 Btu/hr. This is a significant amount of heat. From experience it can be shown that 80,000 Btu/hr is adequate to maintain a not too well insulated tri-level home at an internal temperature of 65°F when the external temperature in Rochester, NY, is −15°F with moderate winds. The waste heat from the system would be adequate to heat a rather substantial home.

This waste heat rejection is necessary in order to turn the exhaust steam back into water for restoration to the boiler. If the engine were operated in a noncondensing fashion and allowed simply to exhaust to the atmosphere, there would be several sequelae, none of them advantageous.

First, if the exhaust pressure were raised to 14.7 psia, we see from the steam tables that the available energy would be 1198 Btu/lb − 1150 Btu/lb = 48 Btu/lb. This is just a bit more than half of the energy attainable with the condenser. We would have to generate nearly twice as much steam.

Secondly, if the engine exhausted to the atmosphere, it would be necessary to provide valves at the top of the cylinder to vent the cylinder to the atmosphere. Otherwise the engine would give up even more of our hard-won solar energy in compressing the residual steam to five, six, or more times atmospheric pressure.

Thirdly, the exhaust noise would be a distinct problem and would require extensive muffling. Those old enough to recall the steam locomotives will adequately attest to the noise of a noncondensing steam engine.

Finally, the boiler water would be lost. Ordinary tap water and lake and river water contain far too much dirt and dissolved minerals to be run through the boiler untreated. On a small system such as this, it is practical to operate the engine on an initial charge of distilled water and to provide a distillation apparatus that can be easily cleaned of boiler scale to provide a continuing supply of makeup water.

The condenser can be designed to be either liquid- or air-cooled. For air-cooled designs, it is common to attain heat transfers of 11 to 14 Btu/ft$^2$/hr/°F. Water-cooled condensers are much more common and much more compact. In water-cooled designs it is common to achieve a cooling rate of 150 to 250 Btu/ft$^2$/hr/°F, with the steam emerging 10° to 20°F above the water outlet temperature and a water temperature rise of 15° to 20°F. To achieve a steam pressure of 1 psia, we would require a water inlet temperature on the order of 70°F, which is within the range of outlet temperatures for a radiant heating system for a home. Presuming that the temperature drop across the condenser tubes was 15°F and that we could attain a transfer rate of 200 Btu/ft$^2$/hr/°F, we would require a condenser tube area of 119,832 Btu/hr divided by 3000 Btu/ft$^2$ = 40 ft$^2$. For ordinary half-inch copper water line, this would require about 270 lineal feet. The water would emerge at a temperature of 85° to 90°F, which would again be within the range capable of providing good radiant heating for a home.

# Back to the storage tank

To provide this action, the boiler would have to provide not only the 10,768 Btu actually utilized in the engine but also the 119,832 Btu rejected by the condenser for a total of 130,600 Btu/hr. As a cross-check we see that the 1198 Btu/lb contained in the 200 psia steam minus the 70 Btu/lb contained in the condenser water gives an addition of 1128 Btu/lb for the boiler. Multiplying this by 115.78 lb/hr yields 130,600 Btu/hr.

If we presume a boiler heat input (with the engine inoperative) of 130,600 Btu/hr and we presume that we would like to have the boiler up-to-steam from a starting temperature of 32°F in perhaps 15 minutes, we may calculate the amount of water in the flash tank. At 381.8°F we see that water has a heat content of 355 Btu/lb. Dividing this into a quarter of our hourly input gives an answer of 92 lb for the flash tank contents.

# The circulation pump

Next let us see what is required of the circulation pump. If we presume that we employ a pressure differential of 50 psi between the boiler and the flash tank to prevent the boiling in the boiler proper, we see from the tables that we could add 21 Btu/lb to the water in the boiler. For the required heat transfer this would mean that we would have to pump 130,600 Btu/hr divided by 21 Btu/lb = 6,219 lb/hr through the boiler. Since water is incompressible, the work required is simply the force times the distance and the power required may be calculated:

$$50 \text{ psi} \times 144 \text{ in}^2/\text{ft}^2 \quad \frac{\times 6219 \text{ lb}}{62.4 \text{ lb/ft}^3}$$

$$= 717{,}582 \text{ ft-lb/sec}$$
$$= 199.33 \quad \text{ft-lb/hr}$$
$$= 0.36 \text{ hp}$$
$$= 270 \text{ watts}$$

The water would undergo a 19.2°F temperature rise.

# The feedwater pump

The feedwater pump has to move the water from the condenser to the boiler against a pressure differential of 200 psi. For this the calculation becomes:

$$200 \text{ psi} \times 144 \text{ in}^2/\text{ft}^2 \quad \frac{\times 115.78 \text{ lb/hr}}{62.4 \text{ lb/ft}^3}$$

$$= 53{,}437 \text{ ft-lb/hr}$$
$$= 14.84 \text{ ft-lb/sec}$$
$$= 0.027 \text{ hp} = 20.4 \text{ watts}$$

The feedwater pump that fulfills the compression portion of the vapor cycle requires a negligible portion of the cycle energy in a vapor cycle engine.

On the other hand it can be seen that the circulation pumping does require a significant amount of the energy. This is one of the disadvantages of a system design which requires that the heat addition take place at a point physically removed from the flash tank. This is probably a penalty difficult to avoid in a tracking solar collector system.

# The dP regulator

From the foregoing discussion of the bimodal operation of the circulation pump, it should be relatively obvious that the item designated dP in the system diagram should not be just a simple constriction in the piping but rather should be a pressure regulator. Figure 15-5 shows one configuration of such a device. The pressure differential between the solar boiler and the flash tank presses against the diaphragm and compresses the spring, thus separating the moving jet from the fixed jet. As the water flow increases, the jets open wider and tend to allow more water to flow, thus holding the pressure differential across the regulator to a relatively constant level. For the idling condition the small low-pressure bypass permits a small amount of flow. The inclusion of this regulator provides for a smoother operation of the system.

# The air valve

The air valve shown at the top of the flash tank serves to admit air to the flash tank whenever the temperature in the flash tank falls below 212°F. This prevents the flash tank from being exposed to atmospheric crushing and water suction from pushing water slugs into the steam piping. At start-up, this valve is held open until

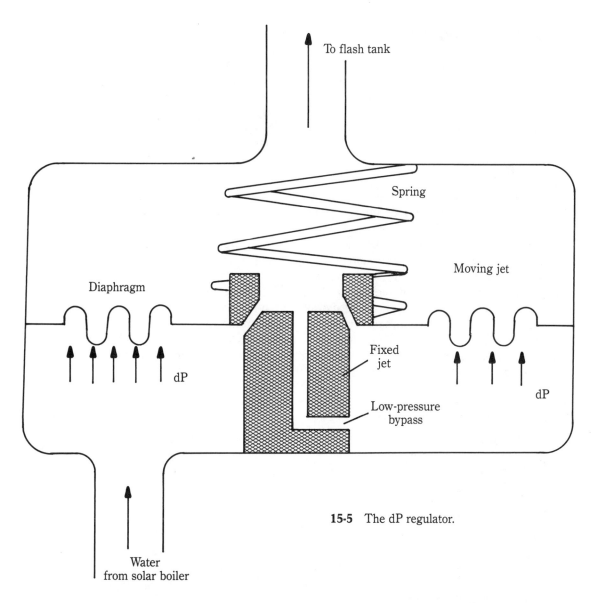

To flash tank

Spring

Moving jet

Diaphragm

Fixed
jet

dP

Low-pressure
bypass

dP

**15-5** The dP regulator.

Water
from solar boiler

the water in the flash tank reaches atmospheric boiling temperature. Once boiling
begins in the flash tank this valve closes. The steam will expel the majority of the
air from the system.

## The vacuum pump

The purpose of the vacuum pump is to remove the air and gasses dissolved in the
boiler fill. This unit operates only periodically. It must be capable of developing a
vacuum equivalent to the best condenser vacuum in order to function effectively.

# The safety valve

The purpose of the safety valve is to limit the pressure within the flash tank to the design level for the system. This valve should be designed following the ASME Boiler Code or some overriding legal code. The code specifies not only the opening pressure but also the size of the valve outlet. The safety valve is a most important part of the system. Without the safety valve, the flash tank is a time bomb!

# General

A few general notes are in order on the subject of the system. First and foremost of these is the fact that the various pumps, tracking motors, and control apparatus should be arranged so that they operate on the auxiliary dc storage power. This is necessary to permit system start-up. The system cannot be operated without these items and therefore cannot depend upon running system output.

Secondly, the system piping should be well insulated on all of the hot lines in order to maintain some reasonable standards of safety and efficiency.

Thirdly, the safety aspects of the design of such a system cannot be overemphasized. At temperatures approaching 400°F, steam can melt human flesh from the bones in seconds. The design of even such a small-scale system requires the exercise of very good mechanical engineering practices and should not be attempted by amateurs.

# Other heat engines

With reference back to chapter 11, we see that any type of heat engine that has the compression, heat addition expansion, and heat rejection cycle could be used. One unit operating on the sterling cycle principle with the heat receptor located at the focus of a small paraboloid is commercially available. However, the size of this unit places it more nearly in the class of a toy than of a practical unit.

In principle, a gas turbine system that compressed air in a rotary compressor, heated it in the solar boiler equivalent, and then expanded it back to the atmosphere or to a condenser equivalent could be made to operate with reasonable efficiency. However, the cost of development of such a unit is prohibitive for the individual. As a practical matter, the use of solar steam appears to be the most practical choice at this writing.

# 16

# Alternate energy storage techniques

AS NOTED IN CHAPTER 4, THE ONLY REALLY ECONOMICALLY VIABLE MEANS of storing any significant quantity of electrical energy at the time of this writing is by means of lead-acid storage cells. Unfortunately, these leave a great deal to be desired in terms of capital investment, maintenance, and life-cycle replacement cost. Despite these factors, the lead-acid battery is still used by telephone companies and many computer installations for standby power because of lack of a viable alternative.

Few of the power companies have any significant means of storing excess energy. Those few that do, accomplish the storage by running a hydro plant in reverse to pump water into the impoundment. Typical overall efficiency for such operations is approximately 60 percent. The pumped water storage is really a very satisfactory system provided there is room for the impoundment. Probably one of the best and most trouble-free home electrical installations would consist of a series of water-pumping windmills feeding an impoundment and a hydroelectric plant drawing water from the impoundment. This alternative is of course available only to those homes with suitable amounts of real estate to contain the impoundment and with ready sources of water. The considerable advantages of a hydro installation were noted earlier.

## Hydrogen

One of the commonly discussed mechanisms for storing of excess electrical energy from a fluctuating source entails the release of hydrogen from water by electrolysis. This is a very efficient process in terms of Btu's of fuel content versus overall electrical input. In addition to this, hydrogen is a nearly ideal fuel. Hydrogen will burn without any soot or residue and if burned with the oxygen also liberated in the electrical process, the only combustion product is water vapor. If the hydrogen is burned in air with a slight excess of air under high temperature and pressure as in an internal combustion engine, there can be some $NO_x$ formation; however, if a slight excess of hydrogen is present, this will not be the case. Hydrogen is also about the most energetic fuel there is, with a heat of combustion of about 52,000 Btu/lb. This is more than 2.5 times the energy of gasoline, 3.5 times the energy of coal, and 7 times the energy of wood.

Unfortunately, aside from the "Hindenberg Syndrome," there are some very real problems associated with the handling and use of hydrogen as a fuel. Hydrogen weighs only 0.0056 lb/ft$^3$ at standard temperature and pressure. It would therefore require a volume of 179 ft$^3$ to contain a pound at standard temperature and pressure. An internal combustion engine that is very efficient will run on about 0.24 lb of hydrogen per horsepower-hour. Our 3-horsepower example would therefore be consuming something on the order of 129 ft$^3$/hr of hydrogen, and a 24-hr run would require about 3096 ft$^3$. The gas could be stored in a balloon; however, extreme caution would have to be exercised to keep all oxygen out of the balloon to avoid a "Hindenberg reaction."

Of course, hydrogen is used as a propellant in some of our highest performance rockets; however, it is used in liquefaction form. There are a few problems with liquefication of hydrogen. At room temperature, liquid hydrogen has a vapor pressure on the order of 10,000 psi. Vessels and plumbing to contain it would have to be immensely strong and tight. On the other hand, the boiling point of liquid hydrogen is –422.9°F. An open vessel of cryogenic liquid hydrogen will maintain itself at this temperature by boiling off slight amounts of the liquid. The space vehicles use hydrogen in this cryogenic form. Unfortunately, this is not all cakes and ale either. The hydrogen which boils off to keep the temperature down must be allowed to escape the vessel. If vented to the atmosphere, it will rapidly rise and dissipate; however, the vicinity of one of the tanks is a definite no-smoking area. Furthermore, most of the common materials become brittle as glass at these low temperatures. It seems likely that cryogenic liquid hydrogen will remain a fuel to be manipulated only by experts in high-technology applications.

Probably the greatest hope for the use of hydrogen as a fuel is the possibility of the development of a solvent-carrier technique in which the hydrogen is dissolved in a fluid the way carbon dioxide is dissolved in soft drinks or beer. Such a technique is currently widely used for acetylene shipment. Experiments have been reported in which hydrogen has been dissolved in lithium hydride and other chemicals, but at this writing no such process or apparatus exists on a commercial scale.

# Wood and vegetation

Probably the easiest means of energy storage available to the home electrifier is simply the growing of wood or vegetation that can then be burned. One of the great advantages of the solar steam installation is that the system can easily be modified to employ a conventional fuel-burning boiler to provide steam for the system when the solar energy is not available. Wood will provide between 7,000 and 8,000 Btu/lb, depending upon the type. Unfortunately, wood and foliage tend to burn in the initial damp phases with a great deal of smoke. One possible solution to this problem would be to make use of the excess steam produced during periods of high solar flux to drive off the moisture from the vegetation or wood.

If the drying is performed in a sealable vessel with the steam tubes on the outside, the process can proceed in three steps. In the first step, the wood or vegetation can be brought up to some temperature like 220°F with the vessel vented to the atmosphere. After a holding period at this temperature, the overwhelming majority of the moisture will be driven off. After this, the vessel may be sealed from the atmosphere and the temperature taken up to near the boiler temperature. During this period, the wood or vegetation will emit a volatile combustible gas. This gas is directly suitable for use as a fuel in the boiler or in an auxiliary internal combustion engine. Depending upon the temperatures achieved in the retort, the wood or vegetation residue in the retort might be entirely or partially converted into charcoal, which makes a fine hot-burning smokeless fire for heating boilers (or for blacksmithing or broiling steak). A portion of the initial drying could of course be accomplished simply by placing the material in racks above the ground exposed to the sunshine. However, except in very dry climates, the drying will not be very complete, and the initial gas produced from the retort will have a large moisture content.

It should be noted that the gasses produced by this process are explosive and the venting should be accomplished through a suitable flame-holder to prevent an explosion in the retort. Unless a tall chimney is available for obtaining a natural draft, the charcoal fire will have to be supplied with a forced draft from a blower if a sufficient fire for steam production is to be maintained. The charcoal product obtained from this process has the advantage that it may be stored nearly indefinitely without loss and is always ready for energy production. Storage should be in a dry place or shelter.

# Direct heat storage

In the previous discussion, the flash tank was considered to be about minimum in size. It has been suggested that a very large flash tank could be employed to store heat directly. This tank could be buried in a large excavation and insulated with some high-temperature insulation material in a very heavy layer. The principle difficulties with this type of scheme are first the problems involved with keeping the insulation dry so that the insulation value will be retained and, second, the cost involved in building a large high-pressure vessel. There is an advantage in scale in schemes of this sort since the surface area of a vessel tends to grow as the square of the radius, whereas the volume of the vessel tends to grow as the cube of the radius. This means that a large vessel will suffer a smaller percentage of heat loss than a small vessel. The required wall thickness also tends to grow as the square of radius, whereas the energy storage capacity tends to grow as the cube of radius. Naturally, a large storage system containing enough energy to permit a coast-down time of several days to a week would come up to temperature slowly as well and might require a time on the order of weeks to a month to get up-to-steam from an initial cold start.

# Compressed air

Another energy storage scheme that has sometimes been suggested makes use of compressed air pumped into a cave or a mine. Aside from the problems of leakage, this scheme suffers from several problems. First of all, the compressors that are used to pump the air in the first place tend to deliver the air hot; that is, a significant portion of the work in compressing the air is contained in the heat. Unless the cave were well insulated, this heat would eventually leak off into the soil and would be lost. Upon emerging from the cave and being expanded, this air would become quite cold in the engine, thus creating a further inefficiency. Of course, the cooling action could be used for refrigeration or air conditioning. The second problem is perhaps more serious. There are a number of locations where there is a small continuous leakage of natural gas from the depths of the earth. In a large cave, this leakage could eventually reach explosive proportions. In addition, it is very difficult to operate a large compressor without the entertainment of a certain amount of oil vapor in the output. For either or both of these causes, the development of an explosive mixture within the compressed air is a distinct possibility. A spontaneous spark due to rocks that fall from the ceiling or to some other cause is another distinct possibility. Unless great care is taken in monitoring the air for explosive vapors or gasses, the possibility of a huge explosion exists.

The molten salt technique described in connection with the power tower is also probably beyond the reach of the individual. In the first place the energy required to melt the salt initially so that it can be pumped through the system is significant. Furthermore, at times when solar input is not available, more energy must be supplied to prevent having the salt solidify within the system, thereby destroying it. Finally, the technology of the construction in a system capable of withstanding the corrosive effects of the molten salt is not child's play.

# Summary

For those having the geographic advantages and the real estate to make it possible, the pumped water energy storage is probably the best available. For the rest of us, the consumption of wood or cumbustible vegetation seems to offer the cheapest and most viable form of energy storage. Future developments in vehicle-solute technology might eventually make hydrogen storage the best and least expensive of all.

# 17

# Summary

THE ENGINEERING STUDY THAT LED TO THE PRODUCTION OF THIS BOOK was intended to develop some basic points:

- A determination of the economic and engineering feasibility of the production of home electric power by the individual.
- Development of a list of practical resources for acquisition of the necessary apparatus.
- Development of a group of design rules for the construction of a practical and effective system.

In keeping with these goals, a series of nearly 70 letters were sent to manufacturers who advertised equipment that might be suitable or could perhaps be modified to be suitable for home electricity production. These letters spelled out in detail the requirement for equipment in the 2- to 25-kVA range and were tailored to the product of the manufacturer. The responses were sparse. A few of the manufacturers of high-speed turbines kindly responded but noted that their products were high-precision items intended to operate fuel pumps on guided missiles, provide propulsion for torpedos, or drive spin testing machinery. The only responses for steam reciprocating engines came from people in the hobby market, which caters to an audience interested in the antique.

A similar lack of response marked the investigation into the small hydraulic turbines. While there have been references to turbines manufactured in Europe, no catalog sheet, price information, or engineering specifications for suitable units were submitted.

In the area of windmills for electrical power generation, there are several units manufactured in the US and in Europe that are capable of producing nearly a kW under high wind conditions; however, the average production for such units in even relatively windy areas (9 mph avg.) falls to such a level that they would be capable of powering a desk lamp and *perhaps* a radio but *not* a color TV.

The conclusion to be drawn from this investigation is that manufacturers have not yet found it economically feasible to address this market. At the time of this writing, it seems that the would-be home electrifier is going to have to build the overwhelming portion of the apparatus himself.

In the areas of dc to ac conversion, and computer load control, the picture is somewhat brighter since commercial equipment is available, albeit expensive. A number of vendors supply these items with suitable ratings. On the other hand, no manufacturer seems to supply automatic power factor correction apparatus.

Throughout the study, the outlook has been maintained that a home electrical system should be capable of doing the entire job and supplying all the electrical needs of the family. A system that will take care of 10 percent of the requirements in a modern home was considered to be trivial because it is expensive to install and would not be economically justifiable: It would only slightly reduce the electric bills. This might be an interesting hobby item and would certainly provide a conversation piece, but it can in no sense be described as practical. As nearly as it is possible for me to determine, there is not a single home in the US that can be described as a modern electrified home, completely independent of the utility companies (excluding diesel motor generators).

Perhaps the most surprising part of the study has been the extent to which Americans have become energy consumers. The picturesque and huge German and Dutch wind-powered sawmills actually produced no more power than a fair sized chain saw that can be lifted with one hand. The big, old waterwheel that ran the flour mill and ground grain for an entire village produced less power than the typical American family uses with only an air conditioner and vacuum cleaner. The maintenance of our standard of living requires a lot of power.

As a result of this, it can be seen in the preceding chapters that only a few people could hope to provide home electrical power at a net economic savings. If one lived atop Mount Washington, where the wind averages 35 mph, a rather modest windmill would suffice to electrify the home.

Similarly, someone living in a mountainous area in the vicinity of a substantial waterfall could have hydroelectric power at an economic saving. The dweller in Arizona or New Mexico or other sun-drenched areas could get by with a very small solar collector on the order of 22 feet in diameter. However, the difficulty of obtaining Fresnel lenses or mirrored reflectors of suitable size and the costs of steam plant installation and maintenance would make this a questionable financial investment.

This entire picture is of course predicated upon the concept that cost of electrical energy from the utility companies will grow no faster than perhaps a half percent higher than the general inflation rate. A faster growth rate in utility prices would of course make the investment in a home electrical generation system more attractive. Time alone will tell in some of these matters. Conceivably, a rising growth rate in utility prices might also bring about a change of zoning requirements so that we might begin to see 22-foot windmills or 32-foot solar reflectors rising above houses on suburban lots. Unfortunately, I fear that neighbors are not ready for them right now.

# Index